MICROTECHNOLOGY AND MEMS

MICROTECHNOLOGY AND MEMS

Series Editor: H. Baltes   H. Fujita   D. Liepmann

The series Microtechnology and MEMS comprises text books, monographs, and state-of-the-art reports in the very active field of microsystems and microtechnology. Written by leading physicists and engineers, the books describe the basic science, device design, and applications. They will appeal to researchers, engineers, and advanced students.

Mechanical Microsensors
By M. Elwenspoek and R. Wiegerink

CMOS Cantilever Sensor Systems
Atomic Force Microscopy and Gas Sensing Applications
By D. Lange, O. Brand, and H. Baltes

Micromachines as Tools for Nanotechnology
Editor: H. Fujita

Modelling of Microfabrication Systems
By R. Nassar and W. Dai

Laser Diode Microsystems
By H. Zappe

Silicon Microchannel Heat Sinks
Theories and Phenomena
By L. Zhang, K.E. Goodson, and T.W. Kenny

Shape Memory Microactuators
By M. Kohl

Force Sensors for Microelectronic Packaging Applications
By J. Schwizer, M. Mayer and O. Brand

Integrated Chemical Microsensor Systems in CMOS Technology
By A. Hierlemann

CCD Image Sensors in Deep-Ultraviolet
Degradation Behavior and Damage Mechanisms
By F.M. Li and A. Nathan

F.M. Li
A. Nathan

CCD Image Sensors in Deep-Ultraviolet

Degradation Behavior
and Damage Mechanisms

With 84 Figures

 Springer

Flora M. Li
Arokia Nathan
Electrical & Computer Engineering
University of Waterloo
Waterloo, Ontario N2L 3G1
Canada
Emails: fmli@uwaterloo.ca
anathan@uwaterloo.ca

Series Editors:

Professor Dr. H. Baltes
ETH Zürich, Physical Electronics Laboratory
ETH Hoenggerberg, HPT-H6, 8093 Zürich, Switzerland

Professor Dr. Hiroyuki Fujita
University of Tokyo, Institute of Industrial Science
4-6-1 Komaba, Meguro-ku, Tokyo 153-8505, Japan

Professor Dr. Dorian Liepmann
University of California, Department of Bioengineering
466 Evans Hall, #1762, Berkeley, CA 94720-1762, USA

ISSN 1439-6599
ISBN 3-540-22680-X Springer Berlin Heidelberg New York

Library of Congress Control Number: 2004116223

This work is subject to copyright. All rights are reserved, whether the whole or part of the material is concerned, specifically the rights of translation, reprinting, reuse of illustrations, recitation, broadcasting, reproduction on microfilm or in any other way, and storage in data banks. Duplication of this publication or parts thereof is permitted only under the provisions of the German Copyright Law of September 9, 1965, in its current version, and permission for use must always be obtained from Springer. Violations are liable to prosecution under the German Copyright Law.

Springer is a part of Springer Science+Business Media

springeronline.com

© Springer-Verlag Berlin Heidelberg 2005
Printed in The Netherlands

The use of general descriptive names, registered names, trademarks, etc. in this publication does not imply, even in the absence of a specific statement, that such names are exempt from the relevant protective laws and regulations and therefore free for general use.

Typesetting: by the authors and TechBooks using a Springer LATEX macro package
Cover concept: eStudio Calamar Steinen
Cover production: *design & production* GmbH, Heidelberg

Printed on acid-free paper   57/3141/jl - 5 4 3 2 1 0

Preface

As the deep-ultraviolet (DUV) laser technology continues to mature, an increasing number of industrial and manufacturing applications are emerging. For example, the new generation of semiconductor inspection systems are being pushed to image at increasingly shorter DUV wavelengths to facilitate inspection of deep sub-micron features in integrated circuits. DUV-sensitive charge-coupled device (CCD) cameras are in demand for these applications. Although CCD cameras that are responsive at DUV wavelengths are now available, their long-term stability is still a major concern. Since the energy of DUV photons is comparable to the band-gap energy of the silicon dioxide (SiO_2), photoreactions can occur in the SiO_2 layer of the CCD image sensor and introduce anomalous behavior.

Given the relative infancy of research in CCD-DUV interactions, publications in this area are somewhat sporadic. This book describes the degradation mechanisms and long-term performance of CCDs. The material presented in this book evolves from a comprehensive literature survey of the scientific research that underpins degradation behavior of CCDs.

Part I begins with an overview of CCD image sensors, and addresses the issues concerning CCD imaging in DUV, along with the common UV enhancement techniques adopted by the industry. Currently, backside-thinned back-illuminated CCD cameras are the leading contender from the standpoint of UV sensitivity. However, such cameras are expensive, and are vulnerable to radiation-induced instability, similar to other CCD designs. To understand the origins of CCD instability in DUV, knowledge of the reliability issues of the silicon-silicon dioxide (Si-SiO_2) system is essential. In Part II, the properties of Si, SiO_2, and Si-SiO_2 interface are described. On irradiation, the Si substrate, the SiO_2 layer, and the Si-SiO_2 interface interact with the incoming photons resulting in damage introduced to the CCD sensor. A discussion of the general effects of radiation is presented in Part III, where we consider the different types of defects and reactions that can arise in the Si-SiO_2 structures. Material interactions with UV radiation are examined in Part IV to identify the sources of instability in CCDs. Part V introduces experimental data that characterize thinned front-illuminated linear CCD sensors, when subjected to F_2 ($\lambda = 157$ nm) excimer laser irradiation. Mechanisms responsible for the DUV-induced degradation behavior in CCD sensors are identified. Because the mechanisms are associated with the Si-SiO$_2$ system, the analysis can also be extended to other silicon-based UV sensor architectures. Potential design optimization techniques to improve the quantum efficiency (QE) and stability of CCD sensors at DUV wavelengths are discussed in Part VI, followed by concluding remarks and recommendations for future research. A better understanding of the mechanisms underlying DUV-induced degradation of CCD sensors can assist in the design and development of new-and-improved DUV-sensitive CCD sensors.

This work was supported by DALSA Corporation, University of Waterloo, and the Natural Sciences and Engineering Research Council (NSERC) of Canada. The authors acknowledge the contributions by Dr. Nixon O at DALSA Corporation, for his support and the numerous stimulating discussions.

Waterloo, ON, Canada *Flora M. Li*
December 2004 *Arokia Nathan*

Contents

1 Introduction 1
    1.1 Motivation: CCD for DUV Imaging 2
    1.2 Outline of the Book 3

Part I CCD Image Sensors

2 Overview of CCD 7
    2.1 Fundamentals of Solid-State Imaging 11
        2.1.1 Absorption of Photons 11
        2.1.2 Charge Collection 13
    2.2 CCD Response and Quantum Efficiency 14
        2.2.1 Spectral Response 14
        2.2.2 Quantum Efficiency (QE) 15
        2.2.3 Pixel Response Non-Uniformity (PRNU) 17
    2.3 Dark Current 17
        2.3.1 Basics of Dark Current 18
        2.3.2 Sources of Dark Current 18
        2.3.3 Dark Signal Non-Uniformity (DSNU) 19
    2.4 Charge Conversion Efficiency (CCE)
        and Output Node 20

**3 CCD Imaging
in the Ultraviolet (UV) Regime** 23
    3.1 Ultraviolet (UV) Spectrum 24
    3.2 Applications of CCD in DUV 27
        3.2.1 Photolithography 27
        3.2.2 Microstructure Generation 29
        3.2.3 Wafer Inspection System 30
        3.2.4 Beam Profiler for Laser System 31
        3.2.5 DUV Microscope 31
        3.2.6 Spectroscopy 32
    3.3 Techniques to Improve UV Sensitivity 33
        3.3.1 Virtual-Phase CCDs and Open-Electrode CCDs 33
        3.3.2 Backside-Thinned Back-Illuminated CCDs 35

VIII   Contents

        3.3.3 UV-Phosphor Coated CCDs 38
        3.3.4 Deep-Depletion CCDs............................ 39
        3.3.5 Other UV Enhancement Techniques 39
    3.4 Challenges of DUV Detection Using CCDs................. 40

Part II Instabilities in Si, SiO_2, and the Si-SiO_2 Interface

4 Silicon 45
    4.1 Optical Properties of Si 45
        4.1.1 Photoelectric Effect and Si 45
    4.2 Defects in Si 48
    4.3 Si Wafers for CCDs.................................... 49

5 Silicon Dioxide 51
    5.1 Basic Properties of SiO_2............................... 51
        5.1.1 Structural Properties of SiO_2 51
        5.1.2 Optical Properties of SiO_2........................ 54
    5.2 Defects in SiO_2....................................... 54
        5.2.1 Physical Nature of Defects in SiO_2 56
        5.2.2 Formation Reactions of Defects in SiO_2 65
        5.2.3 Behavior of SiO_2 Defects in Si-SiO_2 System.......... 67
    5.3 Instability in Si-Based Devices due to Defects in SiO_2 74
        5.3.1 Reliability Issues due to Optically-Active Defects in SiO_2 75
        5.3.2 Reliability Issues due to Electrically-Active Defects in SiO_2 76
        5.3.3 Instability in MOSFETs due to Electrically-Active Defects 77

6 Si-SiO_2 Interface... 81
    6.1 Physical Structure of the Si-SiO_2 Interface................ 81
    6.2 Defects at the Interface 82
        6.2.1 Formation of Interface States 83
        6.2.2 Carrier Exchange at Interface States 86
        6.2.3 Models of Interface Defects 86
        6.2.4 Electrically-Active Defects at the Si-SiO_2 Interface ... 90

Part III Effects of Radiation on the Si-SiO_2 System

7 General Effects of Radiation 95
    7.1 Overview of Radiation Effects on Matter 95

		7.1.1	Displacement Damage	96

            7.1.1 Displacement Damage 96
            7.1.2 Ionization Damage 96
            7.1.3 Interactions of Photons with Matter 98
     7.2 Radiation-Induced Defects in Si, SiO_2,
          and Si-SiO_2 Interface 99
            7.2.1 Radiation-Induced Defects in Si 100
            7.2.2 Radiation-Induced Defects in SiO_2 100
            7.2.3 Radiation-Induced Defects in Si-SiO_2 Interface 102
     7.3 Effects of Radiation
          on Basic Semiconductor Devices 102
            7.3.1 Radiation Effects on MOS Structures 103
            7.3.2 Radiation Effects on Electro-Optical Devices 107
            7.3.3 Annealing of Radiation-Induced Defects 107
            7.3.4 Radiation Hardening............................ 107

8 Effects of Radiation on CCDs............................ 109
     8.1 Overview of the Radiation Damages in CCDs 109
     8.2 Ionization Damage in CCDs 111
            8.2.1 e-h Generation 112
            8.2.2 e-h Recombination and Fractional Yield 112
            8.2.3 Hole Transport................................ 113
            8.2.4 Hole Trapping 113
            8.2.5 Annealing (Detrapping of Holes) 114
            8.2.6 Interface State Creation 114
            8.2.7 Dependence of Ionization Damage
                   on Insulator Properties.......................... 116
            8.2.8 UV Flood 117

Part IV Interaction of UV Radiation with the Si-SiO_2 System

9 UV-Induced Effects in Si 121
     9.1 Photoemission in Si...................................... 121

10 UV Laser Induced Effects in SiO_2 125
     10.1 Overview of UV Laser Induced Effects in SiO_2 126
            10.1.1 Color Center Formation and Induced Absorption 127
            10.1.2 Density Change 128
            10.1.3 Photorefractive Effect.......................... 129
     10.2 Active Defects in DUV 129
            10.2.1 Oxygen-Deficient Centers (ODCs) 130
            10.2.2 E' Centers..................................... 131
            10.2.3 Non-Bridging Oxygen Hole Centers (NBOHCs) 132
     10.3 KrF and ArF Laser Induced Effects in SiO_2................ 133

           10.3.1 Wavelength vs. Rate of Color Center Formation 133
           10.3.2 Induced Absorption
                  due to KrF Excimer Laser Radiation................ 134
           10.3.3 Induced Absorption
                  due to ArF Excimer Laser Radiation................ 135
           10.3.4 Fluctuations in UV-Induced Absorption 136
    10.4 F_2 Laser Induced Effects in SiO_2 138
           10.4.1 F_2 Laser Induced Defect Formation in SiO_2 139
           10.4.2 Dependence of Defect Formation on F_2 Laser Power .. 141
           10.4.3 F_2 Laser Induced Bleaching
                  of the VUV Absorption Edge 145
    10.5 UV-Induced Charging of SiO_2 147
    10.6 Summary of the UV-Induced Effects in SiO_2 150
           10.6.1 Optical 150
           10.6.2 Electrical....................................... 151

11 **UV Laser Induced Effects
    at the Si-SiO$_2$ Interface**................................... 153

Part V Interaction of DUV Radiation with CCD Sensors

12 **CCD Measurements at 157 nm**............................. 157
    12.1 Experiment Description 157
           12.1.1 Laser Setup...................................... 157
           12.1.2 Frontside-Thinned Front-Illuminated CCDs 159
           12.1.3 Laser Exposure Conditions 161
    12.2 Experimental Results 162
           12.2.1 Response Measurement of Sample-A at 157 nm 162
           12.2.2 Response Measurement of Sample-B at 157 nm 164
           12.2.3 Higher Intensity 157 nm Exposure on Sample-A:
                  Accelerated Degradation Testing 166
           12.2.4 Higher Intensity 157 nm Exposure on Sample-B:
                  Dark Current Measurement........................ 168
    12.3 Analysis of DUV Damage Mechanisms 169
           12.3.1 Analysis of 157 nm Response Measurements.......... 170
           12.3.2 Analysis of Dark Current Measurements............. 176
    12.4 Post-157 nm Measurements 177
           12.4.1 Dark Current Response
                  of Sample-A After 157 nm Irradiation 178
           12.4.2 Dark Current Response
                  of Sample-B After 157 nm Irradiation 180
           12.4.3 DUV-Induced Changes in Visible QE of Sample-A 181
           12.4.4 DUV-Induced Changes in Visible QE of Sample-B 182

12.4.5 DUV-Induced Changes in CCE	186
12.5 Response Measurement of Photodiodes at 157 nm	187
12.6 Summary of CCD Behavior at 157 nm	190
12.7 Future Investigations	191

Part VI Concluding Remarks & Future Research

13 Design Optimizations for Future Research 195
    13.1 Optimization Techniques Based
        on UV Photodiodes 196
        13.1.1 Characteristics of Photodiodes in UV 196
        13.1.2 Techniques to Improve the UV Performance 198
    13.2 Optimization Techniques Based
        on DUV Silica Glass 206

14 Concluding Remarks 207
    14.1 Conclusions 207
    14.2 Recommendations 208

Glossary and Definition of Acronyms 211

References 223

Index 229

1 Introduction

As charge-coupled device (CCD) technology matures, its dominance in the digital imaging market for both the consumer and industrial sectors continues to grow [1]. Although commercial digital cameras have a minimal need for ultraviolet (UV) sensitivity, UV-responsive cameras are in demand for industrial and manufacturing applications. Examples include manufacturing inspection systems, UV spectroscopy, and astronomical applications. More recently, the need for CCD imaging has been extended to even shorter wavelengths in the deep-UV (DUV) spectrum. One of the key applications of CCDs in DUV is to facilitate the inspection of deep sub-micron features and defects of wafers and photomasks in semiconductor inspection systems. Therefore, DUV-sensitive CCD cameras with high-speed, high-resolution, and digital imaging capabilities are crucial for these wafer inspection systems.

However, the development of CCD cameras with a high UV responsivity and long-term stability is not straightforward. Conventional CCD image sensor architectures are usually insensitive to UV radiation (10 to 400 nm). This is because CCD sensors are typically built with silicon technology, and the absorption depth of photons in silicon decreases exponentially as the wavelength decreases. This makes the charge collection in the silicon depletion region extremely difficult. In addition, a large portion of the incident UV photons are absorbed by the frontside gate structure for photogate-based CCD sensors. The polycrystalline silicon (abbreviated as polysilicon) gate material in conventional photogate-based CCDs is at least 400 nm thick, whereas its penetration depth for 400 nm radiation is only 2 nm [2]. Thus, the polysilicon gate effectively shields the active regions of the CCD sensor from incident UV radiation, causing a significant reduction in the quantum efficiency (QE). The gate oxide and passivation layers which are composed of SiO_2 can be another source of interference for the incoming UV photons because the SiO_2 becomes absorbent at UV wavelengths. This situation further deteriorates as the wavelength decreases into the DUV spectrum and below it. As the use of CCD sensors becomes more ubiquitous in manufacturing and lithographic applications, it is critical to develop CCD cameras with a reasonable responsivity and stability for the detection of DUV radiation with shorter wavelengths (higher energies).

1.1 Motivation: CCD for DUV Imaging

One of the main driving forces behind developing CCD sensors for DUV imaging is the continually evolving interest for finer resolution in semiconductor manufacturing. The increasing demands for a higher integration density of integrated circuits (ICs) require device fabrication tools that exploit shorter wavelength illuminations for optical lithography. In these manufacturing applications, the very short wavelengths of DUV light allow the fabrication of devices with dimensions in the deep sub-micron regime. Although wafer production systems with the illumination sources of 248 nm and 193 nm are well established, lithography with 157 nm (F_2) fluorine excimer laser is emerging as a viable technology for the post-193 nm era. In fact, in the year of 2000, the semiconductor industry forecasted 157 nm F_2 excimer laser to be the technology of choice for feature sizes in the 100 nm to 70 nm nodes, and anticipated the appearance of F_2 excimer laser in a new generation of optical lithography systems by 2005 [3,4]. 157 nm lithography is attractive for several reasons; the pivotal reason is that it is an extension of the existing optical lithography systems which has longer DUV wavelengths of 248 nm and 193 nm. Therefore, it is possible to adapt the existing manufacturing and wafer-processing infrastructures to 157 nm lithography systems. In addition, optical resolution enhancing techniques (e.g., phase-shifting masks and off-axis illumination) used in the current generation of lithography systems can be extended to the new generation of 157 nm systems. However, with the recent advancements in immersion-based 193 nm lithography tools demonstrating image resolutions down to 45 nm node and below, the introduction of 157 nm lithography systems in production lines will likely be delayed [5–7]. The 193 nm immersion lithography is predicted to become the toolset of choice for the 65 nm and 45 nm nodes on the semiconductor industry roadmap; those nodes are slated to enter production in 2007 and 2009, respectively [7].

The transition to the 157 nm wavelength is expected to encounter challenges comparable to those of earlier shifts in the lithography wavelength from 365 nm (i-line) to 248 nm, and from 248 nm to 193 nm. Some of the critical issues include the availability of the optical materials and coatings, the mask materials and pellicles, a controlled ambience to minimize contamination of the optical surfaces, as well as compatible sensors and detection devices optimized for 157 nm imaging applications [3]. An example of a sensing device commonly found in mask/wafer inspection and monitoring modules of the lithographic manufacturing systems is the CCD camera. For the reliable operation of manufacturing and inspection systems, CCD sensors must maintain a high sensitivity and stability when they are exposed to DUV. Therefore, the performance of CCD sensors in the DUV regime is one of the many factors that shapes the evolution to DUV lithography. With engineers currently developing production technology for the 90 nm and 65 nm nodes in various wafer fabs, there have been real concerns for the first time where these development engineers lack the needed inspection and metrology technologies at the start of their development work. Thus, the industry requires some breakthroughs that solve the natural limitations to conventional CCD and laser scanning inspection technology at DUV wavelengths [8].

In addition to lithography and semiconductor inspection, numerous other applications can benefit from the development of DUV-sensitive CCD imagers.[1] Therefore, CCD cameras that are sensitive in the DUV must be developed to keep up with the advancements in lithography and in other industrial applications.

1.2 Outline of the Book

DUV-sensitive CCD cameras that are fast, responsive, and stable are desirable for a growing number of industrial applications that requires DUV imaging capability. However, conventional CCD cameras respond poorly in DUV because the frontside polysilicon and oxide structures absorb UV photons, and the absorption depth of UV photons in silicon is short. Moreover, high-energy UV photons can cause irreparable damage to conventional CCD cameras. It is possible to eliminate the use of polysilicon layers in certain CCD architectures (e.g., photodiode-based CCDs) so that the UV problems introduced by the polysilicon can be dismissed. However, the presence of oxide (SiO_2) layers is imperative for any silicon-based CCD structure. Thus, consideration must be given to the properties of the SiO_2 layer when CCD sensors are designed for DUV wavelengths.

The use of CCD image sensors at DUV wavelengths has been investigated by a few groups [9, 10]. Although CCD cameras that are responsive in the DUV are now available, their long-term stability is still a major concern. Despite CCD instability issues, there are very few publications in the scientific literature that address the long-term performance, radiation tolerance, or the causes of degradation of CCDs after exposure to DUV wavelengths. These are the motivations for developing new CCD structures that are optimized for DUV imaging applications, and for thoroughly investigating the DUV damage mechanisms in CCDs.

Following the introduction, an overview of CCD image sensors is provided in Part I, which encompasses a discussion on issues concerning CCD imaging in DUV and examples of UV enhancement techniques exercised by the industry. Currently, the backside-thinned back-illuminated CCD camera is the leading contender for offering the highest UV sensitivity; however, such a camera is the most expensive and it suffers from radiation-induced instabilities similar to those experienced by most other CCD structures.

To understand and recognize the origins of CCD instabilities in the DUV, a general knowledge of the reliability issues of the Si-SiO_2 system is essential. Part II is a discussion of the basic properties and defects of silicon (Si), silicon

---

[1] Refer to Chap. 3 for examples of CCD applications in DUV.

dioxide (SiO_2), and Si-SiO_2 interface. On irradiation, the Si substrate, the SiO_2 layer and the Si-SiO_2 interface can interact with incoming photons to subject the CCDs to a variety of effects or damage. Part III discusses the general effects of radiation on Si-SiO_2 system and on CCDs. This provides the background to understand the type of defects and reactions that can arise from radiation in Si-SiO_2 materials. Material interactions with UV radiation are considered in Part IV to build a foundation for studying the causes of the instabilities in CCDs specifically at DUV wavelengths. The findings gathered from the various scientific research areas are assembled to identify the mechanisms of the CCD behavior in DUV.

In addition to understanding instabilities, a new and more economical technique that enhances the DUV sensitivity of frontside-illuminated linear CCDs is introduced. This technique involves thinning the overlying oxide on the top of the imaging area (i.e., the photodiode) of the linescan sensor to reduce the absorption of the DUV photons in the SiO_2 layer. For simplicity, the discussion of thinned front-illuminated sensors here is confined to CCD structures with no polysilicon on the imaging region; namely, photodiode-based linear CCD architecture. As a result, the research focuses on the concerns associated with the DUV-induced effects in the SiO_2 layer, not the issues that can arise from the polysilicon layer. The characterization of this frontside-thinned front-illuminated linear CCD structure at the DUV wavelength of 157 nm is presented in Part V. By using the theories presented in the preceding chapters, the possible causes and origins of the observed CCD behavior in DUV are revealed. A better understanding of the mechanisms behind the DUV-induced degradation of CCD sensors can assist in the design and development of new-and-improved DUV-sensitive CCD sensors in the future. Finally, the significance of the findings of the research in this book are summarized and future research activities are suggested.

Part I

CCD Image Sensors

2 Overview of CCD

CCD is the abbreviation for *charge-coupled device*. CCD image sensors are silicon-based integrated circuits (ICs), consisting of a dense matrix of photodiodes or photogates that operate by converting light energy, in the form of photons, into electronic charges [11]. For example, when an UV, visible or infrared (IR) photon strikes a silicon (Si) atom in or near a CCD photosite, the photon usually produces a free electron and a hole via the photoelectric effect. The primary function of the CCD is to collect the photogenerated electrons in its "potential wells"(or pixels) during the CCD's exposure to radiation, and the hole is then forced away from the potential well and is eventually displaced into the Si substrate. The more light that is incident on a particular pixel, the higher the number of electrons that accumulate on that pixel. By varying the CCD gate voltages, the depth of the potential wells can be modified. This action enables the transfer of the photogenerated electrons across the registers to the "read-out" circuit. The output signal is then transferred to the computer for image regeneration or image processing. Figure 2.1 denotes that a typical CCD sensor consists of a sandwich of semiconductor layers that are overlaid with a network of gates (or electrodes) to control the transfer of the signal charges from the pixels to the read-out circuitry at the output node [11].

The pixels in a CCD sensor can be arranged in various configurations. Figure 2.2 displays two classic CCD architectures: a linear CCD and an area array CCD. A linear (or linescan) CCD sensor consists of a single line of pixels, adjacent to a CCD shift register that is required for the read-out of the charge packets. The isolation between the pixels and the CCD register is achieved by a transfer gate. Typically, the pixels of a linear CCD are formed by the photodiodes. The CCD shift register is composed of a series of MOS (metal-oxide-semiconductor) capacitors arranged closely across the Si wafer so that the charge can be moved from one capacitor to the next as efficiently as possible. The MOS capacitors are formed by depositing a highly conductive layer of polysilicon on top of the insulating layer of silicon dioxide (SiO_2) that covers the Si substrate [11]. Figure 2.1 provides a cross-sectional view of a linescan CCD sensor.

After the integration of the charge carriers in the photodiodes (i.e., the exposure period), the transfer gate is raised to a high voltage to simultaneously

8   2 Overview of CCD

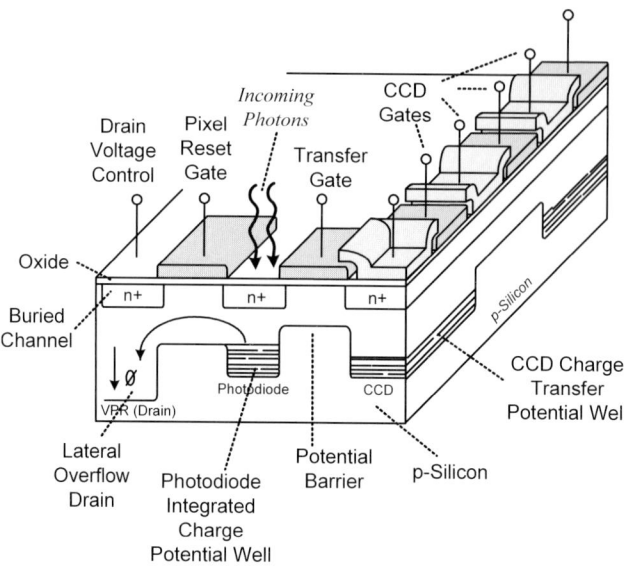

Fig. 2.1. The anatomy of a CCD sensor. The configuration corresponds to a photodiode-based linear CCD (adapted from [11])

transfer the charges in all the pixels in a parallel manner toward the CCD shift register, as indicated by the arrows in Fig. 2.2(a). When the transfer gate voltage returns to its off state, a new integration period begins. At the same time, the charge packets are transferred through the CCD shift register toward the output of the device in a serial fashion. The CCD shift register must be shielded from the incoming light to avoid disturbing the number of carriers in the charge packet [12]. During the imaging operations, the linear CCD sensor is placed along a single axis so that the scanning occurs in only one direction. A line of information from the scene is captured at each integration period, and subsequently read out of the device before stepping to the next line index. An example of linescan imaging is the fax machine. For the purpose of the study in this book, photodiode-based linescan CCDs are used for the DUV experiments in Chap. 12.

Two-dimensional imaging is possible with area array CCD sensors; the entire image is captured with one exposure, eliminating the need for any movement by the sensor or the scene. An example of a compact and simple full-frame area sensor is illustrated in Fig. 2.2(b). The area sensor is composed of an array of photogates (i.e., MOS photocapacitors) that provide both the charge collection and charge transfer functionalities for the CCD sensor. The gate electrode of the photogate is fabricated from polysilicon. A set of parallel light-sensitive CCD registers that is composed of photogates is oriented vertically and is denoted as the vertical CCD registers (VCCD). A simple cross-sectional view of a photogate-based CCD register is depicted

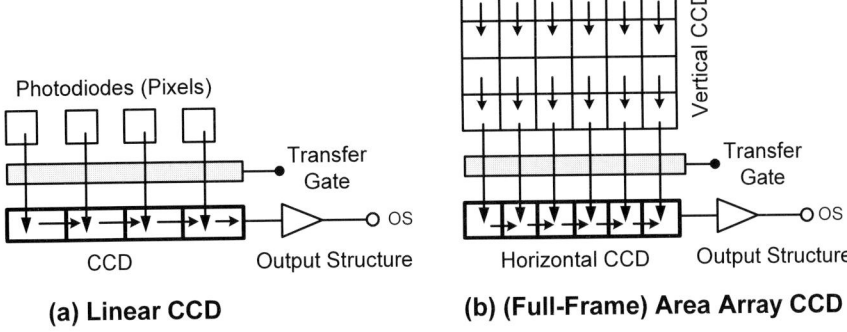

Fig. 2.2. A simplified block diagram of the two fundamental CCD sensor architectures. The arrows denote the movement of the charge packets, facilitated by the clocking signals (adapted from [11])

in Fig. 2.3. At the lower edge of the VCCD array, a single horizontal CCD register (HCCD) is used to combine the outputs from the VCCD into a single output (refer to Fig. 2.2(b)). An additional transfer gate is positioned between the horizontal register and the vertical registers to prevent a charge transfer into the horizontal register, while it is being emptied [11]. During the imaging operations, images are optically projected onto the VCCD array which acts as an image plane. The sensor takes the scene information and partitions the image into discrete elements which are defined by the number of pixels, thus "quantizing" the scene. The resulting rows of the scene information are shifted in a parallel fashion to the serial HCCD register, which subsequently shifts the row of information to the output as a serial stream of data. The process iterates until all the rows of the VCCD are transferred off

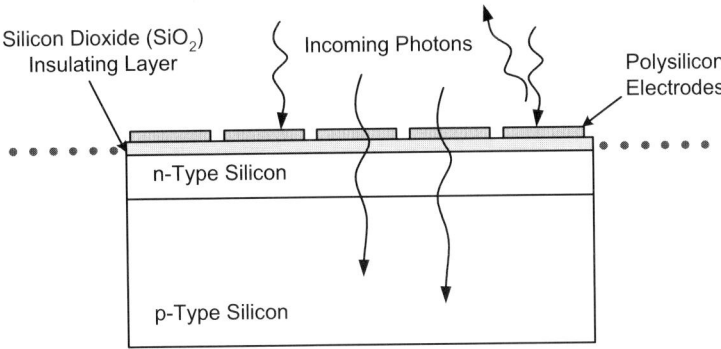

Fig. 2.3. A cross-section of a front-illuminated photogate-based area-array CCD sensor

chip. The image is then reconstructed as dictated by the camera system [13]. Area array CCD sensors are common in consumer digital cameras.

Conventional CCDs are designed for the front-illuminated mode of operation, as depicted in Fig. 2.3. Front-illuminated CCDs are quite economical to manufacture by using standard wafer fabrication procedures, and are popular in consumer imaging applications, as well as industrial-grade applications. However, front-illuminated CCDs are inefficient at short wavelengths (e.g., blue and UV) due to the absorption of photons by the polysilicon layers, if photogates are used as the pixel element. For the front-illuminated area array CCD in Fig. 2.3, the presence of the gate structure (and passivation layers) make it difficult for the short wavelength photons to penetrate the Si substrate, since photons can be absorbed and lost in these upper layers. Furthermore, because of the high absorption coefficient for short wavelength photons in Si and in the overlying materials (e.g., SiO_2, polysilicon), the quantum efficiency (QE) of these front-illuminated CCDs tends to be poor in the blue and UV regions. The gate structure also inhibits the use of an anti-reflective (AR) coating that to boost the QE performance. As a result, the presence of the CCD gates reduces the sensitivity of conventional photogate-based front-illuminated CCDs. Photodiode-based CCDs do not exhibit gate absorption problems due to the absence of the gate (or polysilicon) layers, and offer a higher efficiency at short wavelengths. However, photodiode-based CCDs tend to consume more chip size and may not be suitable for certain applications [13].

Typically, conventional front-illuminated CCDs are adequate for low-end applications and for consumer electronics. But, for large, professional observatories and high-end industrial inspection systems that demand that require extremely sensitive detectors, conventional thick frontside-illuminated chips are rarely used [13]. Thinned back-illuminated CCDs offer a more compatible solution for these applications. Backside-thinned back-illuminated CCDs exhibit a superior responsivity, remarkably in the shorter wavelength region. More details on this design and how it is used to improve UV sensitivity are presented in Sect. 3.3.

Before CCD imaging in UV is examined, the fundamentals of solid-state imaging and the frequently-used terminology for defining the performance of a CCD image sensor are detailed. In particular, QE and dark current are the two key parameters that characterize the CCD behavior in the DUV experiment in Chap. 12. The QE provides a measure of how sensitive the CCD sensor is to incident radiation, and is discussed in Sect. 2.2. Dark current measurements can provide an indication of the extent and type of radiation damage experienced by the sensor. The characteristics and sources of dark current are reviewed in Sect. 2.3. Other parameters such as spectral response, pixel response non-uniformity (PRNU), dark response non-uniformity (DRNU), and charge conversion efficiency (CCE) will also be considered in this chapter; these parameters will be used to evaluate the DUV-induced damages in CCDs in Chap. 12.

2.1 Fundamentals of Solid-State Imaging

The two fundamental components in a solid-state imaging system are the absorption of photons in the device substrate which cause charge generation, and the collection of the resultant photogenerated charge carriers. The effectiveness of these operations is dependent on the optical and electrical properties of the material and the structure of the device. In this section, these two components are described, and their influence on the parameters such as spectral response and QE are considered in the proceeding section.

2.1.1 Absorption of Photons

The first operation of an imager involves the generation of electric charges from the absorption of incident photons. This is illustrated schematically in Fig. 2.4. The charge generation efficiency (CGE) of a CCD is characterized by the QE, and is dependent on the absorption coefficient of the semiconductor material. When the surface of the semiconducting substrate of the imager is struck by a photon flux, Φ_0, the absorption of photons is dependent on the photon energy, E_{ph}, in units of eV, given by

$$E_{ph} = h\upsilon = \frac{h \cdot c}{\lambda} = \frac{1.24}{\lambda_{[\mu m]}}, \qquad (2.1)$$

where h is Planck's constant, υ is the frequency, λ is the wavelength, and c is the speed of light [12]. The photons are absorbed by the semiconductor if the photon energy is higher than the band-gap energy, E_G, of the semiconductor, expressed mathematically as

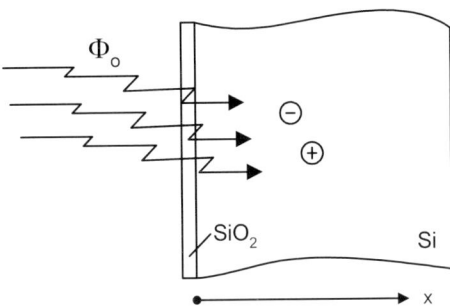

Fig. 2.4. An illustration of the generation of an electron-hole pair due to the impinging photons in the bulk of the silicon (adapted from Theuwissen [12])

$$E_{\mathrm{ph}} \geq E_{\mathrm{G}} , \qquad (2.2)$$

The actual photon flux, $\Phi(x)$, at depth x in the substrate, differs from the incoming flux Φ_0, and is written as

$$\Phi(x) = \Phi_0 e^{-\alpha x} , \qquad (2.3)$$

where α is the absorption coefficient of the substrate material. The values of α of the incident radiation are derived by measuring the absorption intensity, I, of a sample with a thickness, x, in cm, as follows:

$$I(x) = I_0 e^{-\alpha x} , \qquad (2.4)$$

where I_0 is the intensity of incident light exciting the sample [12]. Equations (2.3) and (2.4) are derived from the Beer-Lambert Law. The absorption characteristics of the semiconducting material can also be described by the penetration depth, x^*. It is defined as the depth at which the remaining photon flux, $\Phi(x^*)$, is equal to e^{-1} or 37% of the incoming flux Φ_0, and is expressed as

$$\Phi(x^*) = \Phi_0 \mathrm{e}^{-1} . \qquad (2.5)$$

Thus, the penetration depth and the absorption coefficient appear to have a reciprocal relationship,

$$x^* = \alpha^{-1} . \qquad (2.6)$$

The absorption spectrum of Si (i.e., the dependence of the absorption coefficient, α, and the penetration depth, x^*, on the wavelength, λ) is displayed in Fig. 2.5. An inverse dependence exists between α and λ in the visible-IR

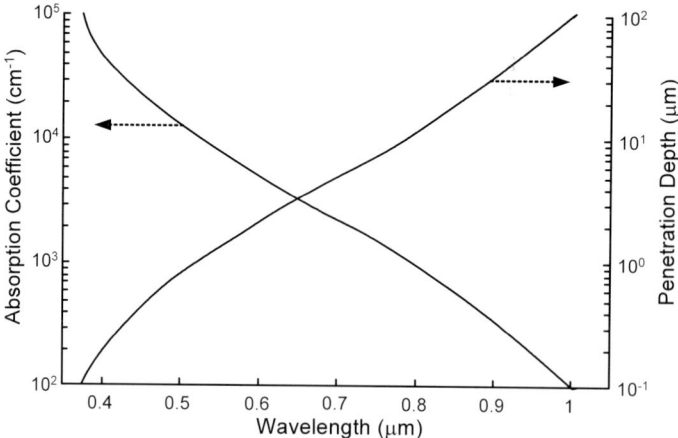

Fig. 2.5. The absorption coefficient and penetration depth of silicon in the visible wavelength spectrum (adapted from Theuwissen [12])

region. This implies that the IR radiation with a longer λ (e.g., 1000 nm) has a smaller α (i.e., a larger x^*) and can penetrate much deeper into the Si before it is absorbed when compared to green light with shorter λ (e.g., 500 nm). For even shorter wavelengths (e.g., blue light), the photon flux is absorbed within a much thinner layer of the Si due to a larger α (i.e., a smaller x^*).

By virtue of the photoelectric effect, the absorption of the photons in the Si results in the generation of electron-hole (e-h) pairs. The amount of charge the is generated depends on the incoming flux, the wavelength of the incoming light, and the absorption coefficient of the semiconducting substrate [12]. For CCD sensors, the charge generation takes place primarily in the Si substrate.

2.1.2 Charge Collection

The other component of a solid-state imaging system is the charge collection. After the generation of the e-h pairs from the absorbed photons, the electrons are separated from the holes within the Si, and collected in the nearest potential well (typically, the depletion region). The CCD cross-section, shown in Fig. 2.1 and Fig. 2.3, contains an n-type Si epitaxial layer on top of a p-type Si substrate, forming a p-n junction that is situated relatively close to the surface of the sensor. An electric field, originating from the depletion region of this p-n junction, causes the e-h pair to separate. All the minority carriers generated in this depletion region are captured by the CCD potential well; thus, the efficiency of charge collection in this region is 100% [12]. However, a portion of the incident photons are absorbed outside the depletion region. For example, some of the photons generate e-h pairs in the neutral bulk of the semiconductor substrate. These carriers in the neutral bulk must diffuse toward the collection site (i.e., the depletion region) in order to contribute to the CCD response. This process of collecting the charges that are generated outside the depletion region relies on the diffusion of the charges to the potential wells; this process has lower efficiency if the diffusion length of the minority carriers is too short, because the charge carriers can be lost by recombination.

The total efficiency of the charge collection process, η_c, for the imager is described as follows:

$$\eta_c = \eta_{dl} + \eta_{bulk}, \qquad (2.7)$$

where the collection in the depletion layer is described by η_{dl} and in the neutral bulk by η_{bulk} [12]. The collection efficiency, η_c, depends on the wavelength of the incident photons and the absorption characteristics of the material. In the case of longer wavelength photons, they have lower absorption coefficients and are likely to have less absorption in the depletion region. As a result, the overall collection efficiency is more heavily dependent on the bulk characteristics of the Si; thus, a longer diffusion length favors a higher collection efficiency for irradiation at long wavelengths. In the case of shorter visible wavelength photons, they are more likely to be absorbed in the depletion region itself, and the recombination characteristics of the bulk material play a less critical role in the collection efficiency. Some photodiode-based CCD arrays have a non-depleted layer above the depletion region and close to the Si-SiO$_2$ interface [12]. The charge generation in this non-depleted region, and the subsequent diffusion to the potential well, must also be accounted for when the charge collection efficiency of the sensor is calculated.

After the charge generation and charge collection, the next step of the CCD imaging operation is the charge transfer. The charges collected at the potential wells are transferred in sequence across the CCD shift registers to the output node. A simple illustration of the charge transfer sequence is depicted in Fig. 2.2. The read-out circuitry converts the charges into electronic signals that can be used to reconstruct an image of the targeted object. The electronic output signal provides a measure for the various CCD performance parameters, including spectral response, QE and dark current, which are discussed next.

2.2 CCD Response and Quantum Efficiency

This section reviews three parameters that are frequently used to characterize the sensitivity of a CCD sensor to the target radiation, which include the spectral response, the QE and the PRNU.

2.2.1 Spectral Response

One parameter that measures the sensitivity of the CCD is the spectral response, R. It provides a measure of the output response of the CCD image sensor due to an optical input signal. R is defined as the ratio of the output current, I_{out}, of the sensor to the incoming light power, Φ_0. I_{out} and Φ_0 have units of amperes per square centimeter (A/cm^2) and amperes per watt (A/W), respectively. I_{out} is the number of minority carriers, Q_n (C/cm^2), collected, divided by the integration time, T_{int}, such that

$$I_{out} = \frac{Q_n}{T_{int}}. \tag{2.8}$$

If the output structure of the CCD sensor uses a floating-diffusion node connected to a source follower[1], the relation between the number of carriers, Q_n, and the output voltage that is measured at the output amplifier, V_{out}, is expressed as

$$Q_n = \frac{C_{FD} \cdot V_{out}}{A_{SF} \cdot A_{cell}}, \tag{2.9}$$

where C_{FD} is the floating-diffusion node capacitance, A_{SF} represents the gain of the source-follower structure, and A_{cell} denotes the area of a single cell or pixel [12].

---

[1] Refer to Sect. 2.4 for the discussion on output structures of a CCD sensor.

2.2 CCD Response and Quantum Efficiency

Combining these parameters in the definition of the spectral response, R, yields

$$R = \frac{I_\text{out}}{\Phi_0} = \frac{C_\text{FD} \cdot V_\text{out}}{A_\text{SF} \cdot A_\text{cell} \cdot T_\text{int} \cdot \Phi_0} . \quad (2.10)$$

Since the absorption of photons and the collection of charges are wavelength-dependent, V_out in (2.10) is wavelength-dependent. Consequently, R is dependent on the wavelength of the incoming light. R also has a high dependence on the characteristics of the substrate material in terms of its diffusion length [12]. If (2.10) is rearranged, the CCD output voltage, V_out, is expressed as

$$V_\text{out} = \frac{R \cdot A_\text{SF} \cdot A_\text{cell} \cdot T_\text{int} \cdot \Phi_0}{C_\text{FD}} . \quad (2.11)$$

V_out is a linear function of the integration time, T_int, and the incoming photon flux, Φ_0. Therefore, there is a larger V_out if the integration time, T_int, is lengthened or if the intensity of the incident radiation, given by Φ_0, is increased. This property of V_out implies that R is linearly related to T_int and Φ_0 at a given wavelength. The relationship of T_int with V_out (and with the dark current) is relevant to the setup and the CCD operating conditions for the DUV experiment described in Chap. 12.

2.2.2 Quantum Efficiency (QE)

In addition to the spectral response, R, a parameter that is commonly used to provide a measure of the sensitivity or responsivity of the device to incident radiation is the QE. It is defined as the number of electrons that are collected, divided by the number of photons incident on the device [12]. This definition is commonly referred to as the "extrinsic" or "external" QE and is a measurable characteristic of a CCD sensor. Not all incident photons can contribute to the QE due to the possible reflection loss at the material surfaces and interfaces, and the absorption loss by the passivation and gate materials. Once the photons are absorbed by the active region of the sensor (e.g., Si substrate), the efficiency of converting the photons to electrons is described by a parameter called the "intrinsic" or "internal" QE. The intrinsic QE is defined as the number of electrons that are generated, divided by the number of photons that are absorbed in the active region of the device. The intrinsic QE is a property or characteristic of the material, not of the device structure [12]. When the QE in real-life applications is discussed, the extrinsic QE is usually implied because it is most frequently used for measuring. The QE is generally a function of the wavelength and temperature.

To derive an expression for the extrinsic QE, it is necessary to consider the energy of the incident photon, E_ph, given in (2.1). The number of electrons that are collected in the CCD is expressed as

$$\text{Number of electrons} = \frac{Q_\text{n}}{q \cdot T_\text{int}} , \quad (2.12)$$

whereas the number of incident photons is represented as

$$\text{Number of photons} = \frac{\Phi_0}{E_{\text{ph}}} = \frac{\Phi_0 \cdot \lambda}{h \cdot c} . \tag{2.13}$$

The QE is a ratio of the number of collected electrons to the number of incident photons,

$$QE = \frac{Q_n \cdot h \cdot c}{q \cdot T_{\text{int}} \cdot \Phi_0 \cdot \lambda} . \tag{2.14}$$

Associating (2.14) with the expression for spectral response, R, in (2.10), R can be defined as a function of QE as follows:

$$R = \frac{QE \cdot q \cdot \lambda}{h \cdot c} . \tag{2.15}$$

Similarly, QE can be expressed as a function of R,

$$QE = \frac{R \cdot h \cdot c}{q \cdot \lambda} . \tag{2.16}$$

Typically, the QE in the visible region (400 nm to 700 nm) peaks close to 650 nm [14]. At wavelengths longer than 650 nm, the sensitivity of the CCD is reduced owing to the small absorption coefficient of the Si (see Fig. 2.5); as a result, it is difficult to absorb photons at these wavelengths in the Si layer. Absorption by the polysilicon gate structure is also a concern. At wavelengths shorter than 650 nm, the absorption in polysilicon increases and becomes particularly pronounced in UV for wavelengths below 300 nm, which degrades the QE for photogate-based CCD structures.

Some CCD designs do not incorporate polysilicon layers on top of the photosensitive region (e.g., photodiode-based CCDs). Absorption loss due to polysilicon is eliminated and these CCDs exhibit improved sensitivity in the visible region. However, their QE still decreases at short wavelengths in the blue and UV regions. This is caused by the short penetration depth in the Si, and the subsequent carrier loss at the Si-SiO$_2$ interface. For example, the penetration depth for 250 nm radiation is only 30 Å in Si, which is only a few atomic layers from the Si-SiO$_2$ interface [14]. The photogenerated carriers can easily become trapped at the interface states and cannot diffuse to the potential well for the charge collection operation. As a result, the QE is lower in blue and UV than the QE in longer visible wavelengths. These characteristics are apparent in the QE curve for the conventional photogate-based front-illuminated CCD in Fig. 2.6. Here, the Si substrate is thinned to 7 μm, and the QE curves are obtained when the CCD sample is illuminated from the front and from the back. The frontside response is limited by the reflection loss and the gate absorption for wavelengths shorter than 550 nm, whereas the backside response is limited only by the reflection loss. For wavelengths longer than 550 nm, the main loss mechanism for both illumination schemes is the transmission loss. The response of the back-illuminated CCDs can be further improved by the application of an anti-reflection (AR) coating [14]. More information on back-illuminated CCDs is provided in Sect. 3.3.2.

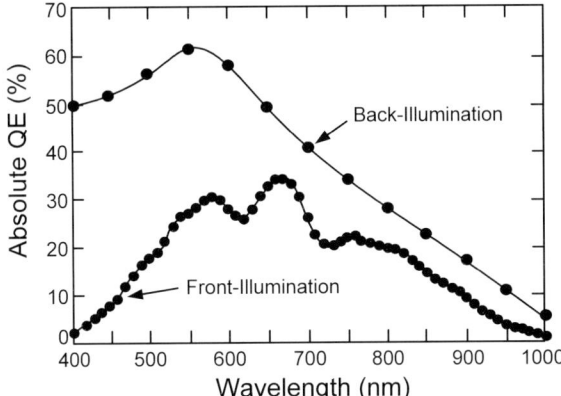

Fig. 2.6. The extrinsic QE performance for a conventional photogate-based CCD sensor with the substrate thinned to 7 μm. The QE curves are obtained when the sensor is operated in the front-illumination mode and in the back-illumination mode (adapted from Janesick [14])

2.2.3 Pixel Response Non-Uniformity (PRNU)

The pixel response non-uniformity (PRNU), also referred to as photo-response non-uniformity, is the pixel-to-pixel variation in the response of a CCD array to a fixed-intensity light, which provides a measure of the QE dispersion over the pixels [12]. Ideally, each pixel in the CCD array has identical response to the illumination. However, due to minor imperfections that are introduced during processing and/or from radiation damage, the pixels in a CCD sensor can exhibit slight deviations in the QE to give rise to PRNU. Usually, the PRNU is wavelength- and temperature-dependent. One approach for evaluating the PRNU involves subdividing the pixels in a CCD array into "windows" or subsections [15]. The PRNU is obtained by calculating the standard deviations of the pixel signals (or QE) in the window. The local PRNU is calculated as the mean of this standard deviation. The global PRNU is the standard deviation of all the pixel signals in the CCD array.

In addition to the parameters discussed above for characterizing the CCD's sensitivity to incident radiation during imaging operations, the noise components of a CCD sensor also play a crucial role in determining the quality of the image produced by the CCD sensor. A very important noise parameter in CCDs is the dark current, and is discussed in the next section.

2.3 Dark Current

One of the commonly observed consequences of radiation damage is the dark current generation. Since the objective of this book is to analyze the effects of

DUV irradiation on CCDs, the dark current can provide a gauge of the UV-induced damages. In this section, the fundamentals and origins of the dark current are reviewed as a preparation for the future discussions on the UV-induced degradation in CCDs.

2.3.1 Basics of Dark Current

The dark current is a source of noise that is intrinsic to semiconductors and naturally occurs due to the thermal generation of minority carriers. The term, dark current, signifies that the measured signal is in the absence of light on the CCD. The dark current is a function of the temperature and is linear with the exposure (or integration) time. At temperatures above absolute zero (0 K), e-h pairs are randomly generated via thermal excitations, and recombine within the Si and at the Si-SiO$_2$ interface [14]. Depending on where the e-h pairs are generated, some electrons are collected in the CCD potential wells and masquerade as signal charges at the output. Large quantities of the dark current limit the useful full-well capacity of the device. Because the generation process is random, the dark current contributes a noise to the CCD output.

Since the dark current is strongly temperature-dependent, the suppression of the dark current is possible by cooling the CCD sensor to very low temperatures. The extent of the required cooling depends largely on the longest integration time that is desired and the minimum acceptable signal-to-noise ratio (SNR). CCDs are most commonly cooled by using a dewar of liquid nitrogen [16].

2.3.2 Sources of Dark Current

There are several principal sources of the dark current. Ranked in the order of importance, they include the charge generation at the Si-SiO$_2$ interface (i.e., surface dark current), the electrons that are generated in the CCD depletion region within the potential well (i.e., depletion dark current), and the electrons that diffuse to the CCD wells from the neutral bulk and channel stop regions (i.e., diffusion dark current and substrate dark current) [14]. These regions of dark carrier generation are identified in Fig. 2.7. In all the cases, the irregularities in the fundamental crystal structure of the Si is responsible for the generation of dark current. For instance, metal impurities (e.g., gold, copper, iron, nickel, and cobalt) and crystal defects (e.g., Si interstitials, oxygen precipitates, stacking faults, and dislocations) are known to be thermal generation sites of charge carriers in Si. These imperfections or impurities within the semiconductor or at the Si-SiO$_2$ interface introduce intermediate energy levels into the forbidden band-gap, which promote dark current generation by acting as "steps" in the transfer of electrons and holes between the conduction and valence bands. This process is also referred to as hopping conduction [14].

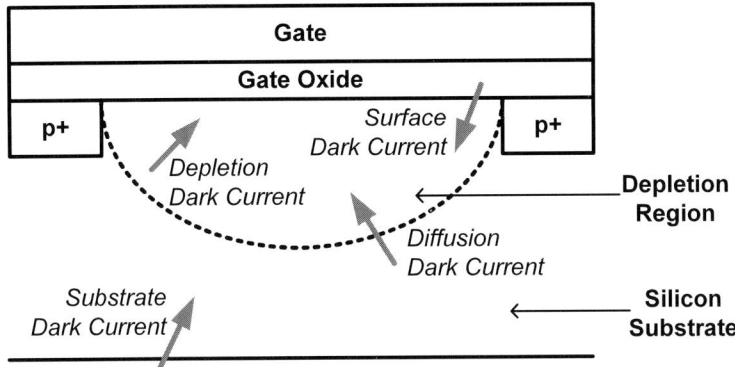

Fig. 2.7. The sources of dark current in a CCD sensor (adapted from Janesick [14])

Of the various dark current sources, surface dark current usually dominates. It is possible to reduce the surface dark current by operating the CCD in inversion mode [14]. This involves applying a sufficiently negative voltage to the gate electrode of a photogate-based sensor, to induce an inversion layer of holes at the interface. The presence of these free carriers fills the interface states, inhibits hopping conduction, and substantially reduces the surface dark current generation rate. With the implementation of the inversion mode, the depletion dark current becomes the only significant source of dark current.

It is also possible that a similar inversion layer can be induced near the Si-SiO$_2$ interface, if the UV-induced damage causes the generation of a sufficient number of negative charges in the SiO$_2$ layer of the CCD. This inversion layer is expected to have a similar influence on the interface trapping dynamics and on the surface dark current; these issues are addressed in Chap. 12.

2.3.3 Dark Signal Non-Uniformity (DSNU)

The dark current generation rate can vary spatially over the CCD array. This occurs because the dark current generation centers are statistically distributed throughout the Si material, so that each pixel contains varying quantities of the generation centers. In addition, the generation rate of each center can vary from type to type. This dispersion of the dark current from pixel to pixel is referred to as the dark signal non-uniformity (DSNU). Some pixels have a very high dark current and are referred to as "dark current spikes" or "hot pixels", which are generally randomly distributed [16]. They result from Si lattice imperfections and impurities, or from lattice damages when the CCD is exposed to high energy radiation sources. The DSNU adds a fixed pattern noise (FPN) to the CCD response signal. Nevertheless, this noise component can be removed from the response signal by image processing techniques. It is also possible to eliminate the DSNU by cooling the CCD [12].

2.4 Charge Conversion Efficiency (CCE) and Output Node

To convert the photogenerated charges or the dark charges that are collected in the CCD pixels into a measurable quantity (e.g., a voltage signal), the CCD sensors employ an output node and a read-out circuitry to perform the necessary functions. The charges that are generated from incident photons are collected in the pixels and transferred across the CCD register to the output sense node. These charge packets are too small to be transferred directly to the outside world without degrading the signal-to-noise ratio (SNR). Thus, an output structure is used to convert the signal charge packets to a voltage which can be amplified before transmission. The performance of this conversion stage is evaluated in terms of the charge conversion efficiency (CCE) which is expressed as the voltage generated per unit charge. There are two common output structures: the floating diffusion output with reset and the floating gate output without reset. In both cases, the converting medium is buffered to the outside world by means of a source follower amplifier [12].

The most widely used output structure is the floating diffusion output with reset. It is easier to fabricate and is less temperature-sensitive. The charge packet that is to be sensed or converted is dumped on a capacitor defined by an n+ floating diffusion region, and the charges in the capacitor are then processed as a voltage signal. After the measurement of each charge packet, the n+ diffusion region is reset by draining away any residual charges; this ensures that the residual charges do not disturb the measurement of the next charge packet. The reset operation is accomplished by connecting the n+ floating diffusion node to a positive supply voltage via a reset transistor; here, the reset transistor acts as a switch and is controlled by a reset clock pulse. The voltage on the floating diffusion output is sensed by a source-follower amplifier stage, which is then fed to the outside world. This configuration is illustrated in Fig. 2.8. The voltage swing at the floating diffusion node for a given charge content is determined by the capacitance of the diffusion node [12]. In some CCD designs, the n+ diffusion region is not completely covered by an overlying gate or electrode to act as a light-shield; as a result, the uncovered region is susceptible to radiation damage. Radiation damage can cause changes in the capacitance and affect the CCE; these changes are believed to be due to the radiation-induced charging in the SiO_2 layer, and is discussed in Chap. 12.

One major drawback of the floating diffusion output stage is its destructive nature of the read-out process. After the voltage on the floating diffusion node has been sensed, a reset operation is performed in every read-out cycle to drain off the residual electrons in the floating diffusion node to the reference power supply. Each reset operation is susceptible to thermal noise which appears as a pixel-to-pixel variation, and gives rise to reset noise that can degrade the performance or efficiency of the CCD. The effect of the destructive read-out can be avoided by using a floating gate output structure, where the

2.4 Charge Conversion Efficiency (CCE) and Output Node

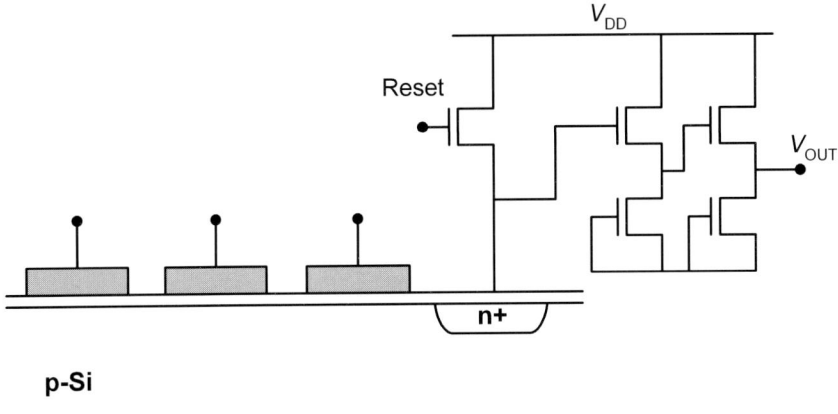

Fig. 2.8. The floating diffusion output structure with reset, for measuring the content of the charge packet transported through a CCD (adapted from [11])

charge packet is dumped into a storage capacitor underneath a floating gate. The packet of charge carriers induces a voltage change on the floating gate, which is sensed by a source follower and fed to the output node. This design is non-destructive because after the sensing operation, the charge packet can be transferred and clocked further through the CCD and sensed once again at another stage, or manipulated by other means. Conventional practice is the use of a floating diffusion sense node, followed by a source-follower amplifier for the output circuitry. Source followers are adopted to preserve the linear relationship between the incident light (or photon flux), electrons generated, and voltage output [12].

3 CCD Imaging
in the Ultraviolet (UV) Regime

In the visible region (400 nm to 700 nm), optical components and CCD sensors are available that offer superb performance. However, for the detection of energetic particles which have wavelengths outside the visible spectrum (i.e., nonvisible imaging), classic solid-state imagers typically have a very low sensitivity in these applications. Figure 3.1 illustrates the various wavelength regions that contribute to the electromagnetic (EM) radiation spectrum. For Si-based solid-state imaging devices, nonvisible imaging is commonly required for the following three wavelength regions: the infrared (IR), the ultraviolet (UV), and the X-ray regions. The near infrared (near-IR) to IR region is described by wavelengths that are longer than 750 nm. For these wavelengths, the corresponding absorption coefficient in Si is too low and almost all the photons pass through the structure without being absorbed. The UV region constitutes wavelengths that are shorter than 400 nm and longer than 10 nm. The absorption in this region is so high that the photons are absorbed in the top layers above the CCD substrate. X-ray refers to the wavelengths that are shorter than 10 nm. For these very short wavelengths, the absorption coefficient of Si is again comparable to that of visible light. However, the energy of the X-ray particles is so high that they can damage the solid-state detector.

The absorption characteristics of Si in the aforementioned wavelength regions are displayed in Fig. 3.2. Despite the complications of the photon

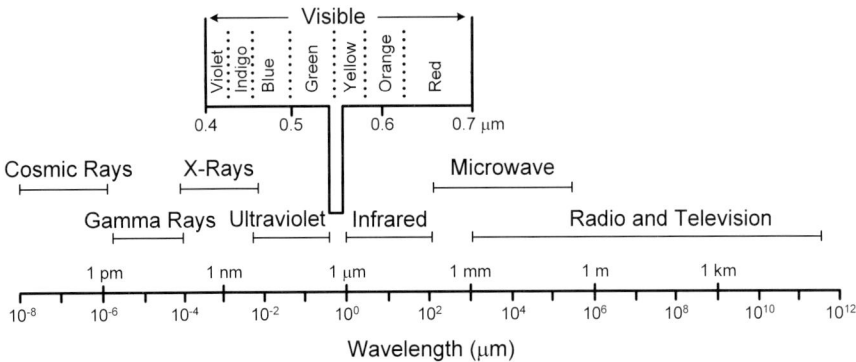

Fig. 3.1. The electromagnetic spectrum (adapted from [17])

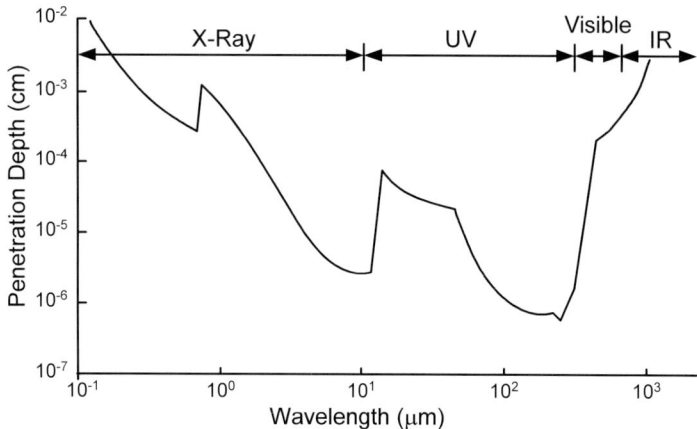

Fig. 3.2. The penetration depth in silicon for the different ranges of X-ray, UV, visible, and IR wavelengths (adapted from Theuwissen [12])

penetration at these nonvisible wavelengths, Si detectors have been designed and tailored to the specific application needs in the X-ray and IR regions. Comparatively, there has been much less effort in the past to develop CCDs to operate in the UV and deep-UV (DUV) regimes. As new applications involving UV and DUV radiation sources continue to emerge, UV-sensitive imagers and detectors are essential. In this section, the characteristics of the UV spectrum and the typical applications of CCD imagers in DUV are reviewed. In addition, the techniques that have been reported for improving the CCD's UV sensitivity and the related challenges for designing CCDs for DUV imaging are highlighted.

3.1 Ultraviolet (UV) Spectrum

Ultraviolet (UV) light can be defined as the small portion (10 nm to 400 nm wavelengths) of the electromagnetic spectrum between the longer wavelength visible light and the shorter wavelength X-ray energy. The term means *beyond violet*, where *ultra* is the Latin word for *beyond*, and *violet* is the color with the shortest wavelength of visible light. Within the boundaries of the UV region, several subdivisions can be identified: extreme, far or vacuum, deep, mid, and near UV. Each subdivision has a particular significance in terms of applications and properties [18].[1]

Extreme UV (EUV) covers wavelengths from 10 nm to 100 nm and is associated with high photon energies. The materials selected for EUV applications

---

[1] The boundaries of the subdivisions of the UV region are often vaguely defined. The definition of the wavelength regions can vary depending on the application and/or the publication.

must exhibit a high damage resistance, high purity, and be free of occlusions or defects that will become energy nucleation sites. *Far UV* or *Vacuum UV (VUV)* includes wavelengths from 100 nm to 200 nm. At these very short wavelengths, the air becomes opaque. Thus, VUV experiments and operations need to be performed in a vacuum (or inert gas) environment so that air does not absorb all VUV light. In commercial UV optical delivery systems, argon- or nitrogen-purged beam containment tubes are used to contain the VUV radiation in vacuum condition. If this is not provided, considerable energy losses will occur due to the absorption of VUV photons by the air molecules. *Deep-UV (DUV)* constitutes wavelengths from approximately 180 nm to 280 nm. This term is related to the fact that DUV is the deepest area of UV, where practical UV imaging is routinely done. It is also the deepest part of the UV spectrum where work can be done in atmospheric conditions without side effects. Although the wavelength of 157 nm requires operation in vacuum, 157 nm is often considered as DUV in the semiconductor lithography industry and belongs to the family of DUV lithographic wavelengths, consisting of 248 nm, 193 nm and 157 nm. *Mid-UV* is the region of the UV spectrum from 280 nm to 315 nm. The name is derived from that fact it is midway between deep-UV and near-UV. *Near-UV* is the region from 315 nm to 400 nm. It is closest to the visible portion of the electromagnetic spectrum [18].

Recently, UV light has attracted increasing attention, as the various technologies that are necessary to provide practical UV laser imaging and beam delivery systems have progressed, with considerable development in UV light sources. Historically, UV light has been available only from relatively low power lamps, thereby restricting the usefulness of this type of radiation. The discovery and development of the excimer laser in the 1980s facilitated the availability of intense UV radiation. As various phenomena involving UV energy and material interactions have been discovered and optimized, practical applications for UV laser have emerged [18]. With characteristics which can precisely remove micrometer-thick layers of tissues or thin films and to vaporize the most refractory materials, UV lasers are indispensable tools in many areas of materials science. Moreover, UV laser radiation offers the capability for the deposition, doping, and modification of semiconductors [19].

For the semiconductor processing industry, the present interest is in DUV and VUV excimer lasers that provide the required high intensity radiation to define sub-micrometer patterns for IC fabrication in photolithography systems. Excimer lasers are a family of pulsed lasers, operating in the UV region. The source of emission is a fast electrical discharge in a high pressure mixture of a rare gas (krypton, argon, or xenon) and a halogen gas (fluorine or hydrogen chloride). The particular combination of the rare gas and halogen gas determines the output wavelength [18]. The commercially available excimer laser systems and their respective output wavelengths are listed in Table 3.1.

Table 3.1. Common UV excimer laser systems [18]

Excimer Laser System	Wavelength	Photon Energy
Fluorine (F_2)	157 nm	7.9 eV
Argon Fluoride (ArF)	193 nm	6.4 eV
Krypton Fluoride (KrF)	248 nm	5.0 eV
Xenon Chloride (XeCl)	308 nm	4.0 eV
Xenon Fluoride (XeF)	351 nm	3.5 eV

The outputs of the excimer lasers typically have pulse energies in the millijoules range, and repetition rates of ten to a few hundred hertz (up to 1 kHz are available). The average power of the excimer lasers is usually in the 10 W to 100 W range. The pulse length is short, typically in the 10 ns to 30 ns range. This results in peak powers of several tens of megawatts [18]. The combination of short UV wavelengths with high peak power and average power, and the temporal and spatial coherence renders excimer lasers to be uniquely suited for a wide range of applications. A summary of fields where UV excimer laser technology is being applied is provided in Table 3.2. Some of these applications require DUV imaging capabilities for process control, and inspection or detection purposes. The next section highlights the use of CCD cameras for various DUV-related applications.

Table 3.2. Applications of UV excimer lasers [18]

Field	Applications
Materials and semiconductor processing	Photolithography, micromachining, chemical vapor deposition, semiconductor annealing, fabrication of high-temperature superconducting films, high resolution machining of plastics and metals, IC packaging (ablation of polymers), cleaning (ablation of particle and film contamination)
Medicine	Corneal surgery, microsurgery, laser angioplasty, lens (intraocular) manufacturing, neurosurgery, bone machining
Environmental science	Fluorescence studies, remote sensing, ozone monitoring
Optical communication	Waveguide fabrication, fiber networks (microstructuring of glass and plastic)
Other	Pumps for high power tunable dye lasers, plasma studies, raman shifting, VUV generation, photolysis

3.2 Applications of CCD in DUV

UV laser technology is evolving from the research phase to commercial phase in a variety of fields. Such technology provides unique functional features such as high resolution, athermal processing, and precise microstructuring and micropatterning capabilities. High power laser sources of DUV radiation play an important role in numerous applications such as microlithography, micromachining, and material modification. Recent progresses in DUV optical materials, laser resonators, and electronics enable dramatic improvements in the power levels, brightness, intensity uniformity, long term stability, and reliability of the DUV excimer laser systems. Since an increasing number of industrial applications are shifting to intense DUV sources, this trend necessitates the development of high performance DUV sensors for process control and monitoring purposes. Examples of DUV imaging applications include semiconductor lithography systems, micromachining, wafer inspection, UV spectroscopy, microscopy, water treatment, and flame monitoring. Electronic imaging with CCD cameras are preferable in these applications, because the CCD technology is mature and it offers a digital imaging solution with high speed, high resolution and low noise capabilities. The electronic image produced by the CCD is available for visualization, interpretation and analysis by the computer. Consequently, there is an interest in extending the CCD technology for imaging at DUV, as well as VUV, wavelengths.

3.2.1 Photolithography

The gate density in silicon ICs quadruples every three years, a phenomenon that is described by Moore's Law. The industry standard for IC fabrication involves optical projection lithography by using a step-and-repeat machine (known as a stepper). The ultimate achievable feature size, and thus the gate density, is limited by the wavelength of the radiation which is used to illuminate the mask which, in turn, projects the desired circuit image onto the Si wafer. The image resolution, or minimum feature size, is proportional to the illumination wavelength. In the pursuit of ever-decreasing feature sizes, it has been necessary for the semiconductor industry to move from the operation of mercury UV lamps at the 436 nm (g-line) wavelength in the 1980s for the linewidth of 0.5 µm, to 365 nm (i-line) in the 1990s for the linewidth of 0.35 µm, and then to a shorter wavelength of 248 nm by the use of KrF excimer lasers to produce feature sizes of 0.25 µm. The incorporation of various resolution enhancing techniques (RETs), including off-axis illumination, various flavors of phase-shifting masks, and optical proximity correction strategies, has enabled the extension of 248 nm lithography to produce the linewidth of 0.18 µm [20]. Steppers utilizing 193 nm light from ArF excimer lasers have been introduced in the fabrication facilities to produce the linewidth of 0.13 µm in 2001. In the late 1990s, the IC industry designated the 157 nm light from F_2 excimer laser as the route to realize device structures at the 100 nm and 70 nm nodes, which was targeted to be in production by 2005 [3, 4, 20]. However, the 157 nm technology was not ready in time for the development of the 100 nm node, and thus, the 193 nm tools are adopted to fill the gap [21]. Furthermore, leading chip makers, like Intel Corporation, had revised their roadmap and planned to use the 193 nm lithography tools to develop IC products based on the 90 nm node starting in late 2003 or early 2004, the 157 nm tools for the 65 nm node in 2005, and finally, the EUV systems for the 45 nm node in 2007 and beyond [22]. Figure 3.3 shows the predicted trend in feature size scaling and lithography wavelength reduction (based on data collected in 2001).

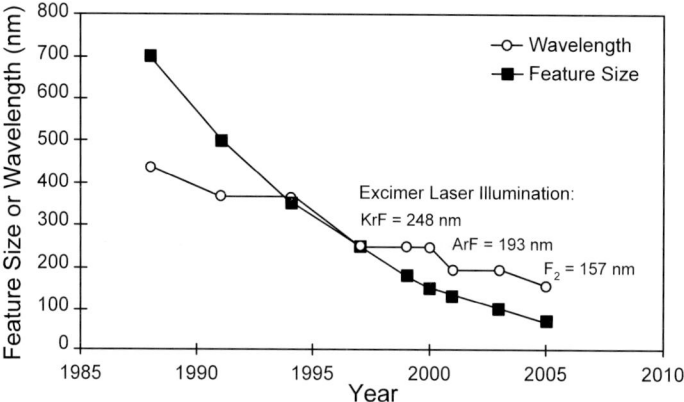

Fig. 3.3. A comparison of feature size and illumination wavelength for semiconductor lithography (adapted from Lalovic [23])

At present, technical issues and delays with the 157 nm steppers in the market are forcing the lithography roadmap to be revised once again. Industrial efforts are focusing on extending the 193 nm lithography tools down to the 65 nm node by 2005 [22]. Furthermore, the recent advancement in the immersion lithography with 193 nm suggests that this technology is likely to extend through the 65 nm node, and perhaps to 45 nm or 32 nm nodes with more extensive research. The essence of immersion lithography lies in using water to change the wavelength of light from 193 nm to 134 nm at the wafer level. The higher refractive index of water at 1.44, compared to air in standard lithography system, allows a higher angle to be sustained in the resist, thereby delivering higher depth of focus (better process window) at a given numerical aperture [24]. If the immersion technology succeeds, then it may be applicable to 157 nm lithography, to extend the wavelength down to 115 nm for realizing even finer feature sizes in the future.

Some of unresolved technical issues in 157 nm lithography, causing its delay for deployment, include the challenges in obtaining and growing the required calcium-fluoride materials for the lithography lens, the high levels of intrinsic birefringence of the calcium-fluoride crystals used to make the lenses, the difficulties in developing suitable resists for 157 nm, and the lack of an acceptable soft pellicle technology. The organic soft pellicles, needed to protect the masks from contamination, tend to degrade and lose their transparency while being irradiated at 157 nm [22].

Another pressing problem in the development of DUV steppers is the need for UV optics and UV sensors with a high performance, that can operate effectively for up to 10^7–10^8 laser pulses at relatively high illumination levels in the DUV and VUV regions of the spectrum (100 nm to 300 nm). Such devices are required for control purposes such as measuring the total UV exposure given to the photoresist covering the Si substrate, and monitoring the laser pulse energy and the spatial distribution of the laser radiation. Conventional Si-based CCDs are currently used for monitoring of mercury UV lamps (g- and i-line). More specialized CCDs have been designed for imaging at 248 nm, but they exhibit unacceptable loss of performance after only 10^5 laser pulses in some applications [25]. Moreover, it is difficult for photons to penetrate the active regions of Si devices at 193 nm and 157 nm, resulting in a very low UV sensitivity for Si detectors and imagers. Other shortcomings of Si detectors in DUV-VUV include a poor radiation hardness and a limited operational lifetime, when intense DUV or VUV radiation sources are being monitored. New CCD sensor architectures with a high DUV-VUV sensitivity and stability must be developed to ensure their compatibility with the upcoming generations of optical lithography systems, and to accommodate the other growing industrial needs for DUV sensors (to be discussed next).

3.2.2 Microstructure Generation

In addition to large-scale lithography systems, DUV excimer lasers are used for microstructure generation in smaller systems, where CCD cameras are employed for sample viewing and process control. These microstructure generation systems involve the projection of a demagnified image of a mask or aperture onto a work-piece at a fluence that is sufficiently high to produce etching. An additional requirement is that the size of the irradiated field be as large as several tens of mm^2. For this area, the beam must be uniform within ±2% and the exposure reproducible within about ±1% on a pulse-to-pulse basis [19]. For the highest spatial resolution and enhanced photo-chemical etching, DUV laser wavelengths of 248 nm, 193 nm, and 157 nm are preferred. CCD cameras are incorporated into the system to monitor beam uniformity and output energy, and to accomplish other processing control tasks. Figure 3.4 signifies the basic elements of an excimer laser-based projector system for microstructure generation. Resolutions, as low as 0.25 µm, have been reported for such a system using a KrF excimer laser [19].

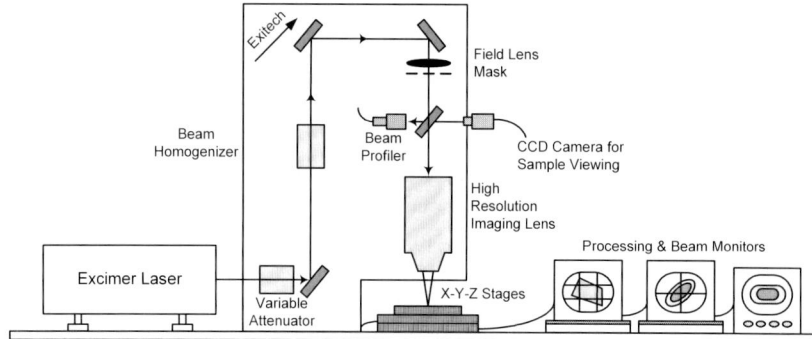

Fig. 3.4. An excimer laser mask macroprojector (adapted from [19])

3.2.3 Wafer Inspection System

As the design rules and process windows continue to shrink, IC manufacturers faced many challenges in achieving and maintaining chip yield and profitability, while moving to new processes such as copper and low-k dielectrics. The incorporation of new materials, new technologies, and new design rules further complicates the goal of maintaining yield. The fabrication facilities must capture a wider range of problems on patterned and unpatterned wafers such as physical defects, electrical defects, and macro defects, which can ruin an entire wafer rather than just a die. Additionally, defects that are not relevant in the older, larger design rules have now become yield-killers at $0.13\,\mu m$ geometries and below. The key to chip yield lies in capturing all these yield-limiting defects while reducing the time to detection, time to correction, and ultimately, time to market [26]. Wafer inspection systems help semiconductor manufacturers to increase and maintain chip yields, by detecting defects that occur during the manufacturing process.

The world-wide semiconductor manufacturing community has relied on cameras which use visible light to inspect the proliferating number of features on photomasks and wafers, and to relay the defect signals to computers. Digital imaging technologies support the tight quality control that is vital to the efficient manufacturing of the silicon ICs that drive today's microelectronic devices. Incorporating CCD cameras in these semiconductor inspection systems allows the detection of flaws and enables proper alignment, increasing yield and throughput. However, as IC fabrication progresses quickly toward smaller critical dimensions, higher chip density, and larger wafers, visible light is no longer practical [27]. In order to facilitate the inspection of deep submicron features on wafers and masks, and to accommodate the increasing throughput requirements, the new generation of inspection systems must follow the lead of lithography and migrate to increasingly lower UV wavelengths for attaining finer resolutions. DUV-sensitive CCD cameras are an asset to these semiconductor inspection systems, since they offer a high speed and high resolution imaging solution. CCD camera with DUV imaging capability is also beneficial in other applications, including beam profiling in laser systems, DUV microscope and spectroscopy, as discussed below.

3.2.4 Beam Profiler for Laser System

Capable of providing the most intense sources of UV radiation, excimer lasers have found a rapidly increasing acceptance in a wide range of applications. For quality control and process optimization, laser beam profiling is fundamental to any laser-related application and laser system (e.g., laser materials processing). Beam profile measurements provide a quantitative output of the intensity throughout the beam area and beam path at a sampling rate that is fast compared to that of the process variables [19]. A simple beam profiling technique involves the use of a CCD camera to directly observe the laser beam. An example of the use of a CCD camera in laser profiling is Lambda Physik's ArF (193 nm) NovaLine laser system. The camera in the NovaLine Lasers is used in conjunction with an etalon spectrometer (monitor module) or grating spectrometer (wavemeter) to determine the wavelength and bandwidth of the laser radiation [28]. In these applications, the attenuated laser is intercepted directly by the CCD. For accurate measurements, proper technique and setup must be executed to ensure that reflections of the beam do not end up on the CCD. This is usually accomplished by using a wedged window, tilting the CCD, and utilizing light baffles and light-absorbing materials inside the camera's vacuum chamber [29]. The advantage of CCD cameras over other electronic beam profiling techniques, such as thermal paper and metalized blocks, is that the CCD provides a high resolution, high accuracy, and quantitative beam intensity information. Also, the use of video CCDs in the laser beam profiling module is now relatively common in many laser laboratories. The benefits of cooled, low-noise CCD cameras are an increased dynamic range, linear response, and improved noise performance.

3.2.5 DUV Microscope

CCD cameras are incorporated into DUV microscopes for small-scale semiconductor inspection and for the defect classification of patterned Si wafers to enhance chip yield. For example, Kubota et al. have reported the use of a 1.4 million pixel CCD camera, specially designed for imaging at the wavelength of 266 nm, in a DUV microscope [30]. In this system, the DUV laser light is focused by a high numerical aperture objective lens, and produces Airy disks with radii as small as 0.173 μm for a 266 nm illumination and 0.130 μm for a 195 nm illumination. Thus, the microscope is characterized by cutoff frequencies of 77 nm L&S (lines and spaces) and 55 nm L&S, for the 266 nm system and the 195 nm system, respectively. These resolutions are compatible with the defect rules of 90 nm and 65 nm for the 0.18 μm and

0.13 μm CMOS processes, respectively. With the incorporation of the CCD cameras, the DUV microscope system is capable of producing high contrast image of defects, which enhances the defect classification capability of the semiconductor inspection tool [30].

3.2.6 Spectroscopy

Spectroscopy is based on the detection of radiation that is emitted or absorbed during atomic fluorescence, and is an important tool for chemical analysis. Spectrometers are used to analyze UV, visible, and IR radiation, and are incorporated in radiometric, fluorometric, luminescence, emission, and Raman systems. Traditionally, spectroscopy is performed by a scanning spectrometer, which involves a scanning monochromator and a single element detector such as a Si photodiode or a photomultiplier tube (PMT) that is placed at the exit slit. In such a scanning system, a complete spectrum is constructed point by point, by moving the grating to select each wavelength [31].

In contrast, in multichannel spectroscopy, the grating is fixed; the exit slit and single detector are replaced with an array of detectors, each detector viewing a different wavelength. This array or multichannel implementation is very efficient compared to traditional scanning spectrometer, because a complete spectrum can be recorded in the same time that it takes to record one wavelength point with a scanning system. Also, array based spectrometers have the potentiality for better stability and reproducibility, since they have no moving parts. Two main arrays are used for multichannel spectroscopy: Si photodiode arrays (PDAs), and charge coupled devices (CCDs). CCDs are at least 100 times more sensitive than PDAs, making CCDs the detector of choice for applications involving low light levels such as Raman scattering and low quantum yield luminescence spectroscopy. Area-array CCDs are used for 2-D spectroscopy. Typically, linear CCDs with 1-D arrays are adopted for spectroscopic applications where there are enough signals, and for applications such as beam profiling and monitoring in which the read-out speed is critical. PDAs have very high saturation levels, which provide ten times the improvement in maximum signal-to-noise ratio than that of CCDs. PDAs are the better choice for reflectance, transmittance, absorbance, and radiometric measurements where high light levels are available [31].

Another key feature of using CCDs in UV spectroscopic microscopy is that the electronic images produced by CCDs can be conveniently visualized, processed, interpreted and analyzed by the computer. Broadband (UV to visible) spectroscopic microscopes uses CCD cameras to take spectrally-resolved transmission and reflectance images, which can be utilized to identify the chemical species in microscopic specimens, and to probe for individual structures for examining the kinetic processes in biological materials. These systems offer a vast range of possibilities for microanalytical imaging. [32].

As spectrometers become smaller and are fully-integrated, there is a need for a sensor that is responsive to a wide range of UV wavelengths. CCD spectrometers, capable of detecting near-UV to VUV radiation from space, are used by astronomers. UV Resonance Raman spectroscopy, used for the analysis of complex biological systems, also requires UV-sensitive CCDs [33].

3.3 Techniques to Improve UV Sensitivity

One primary factor, affecting the performance of solid-state imagers in UV, is the high absorption of UV photons in the Si, SiO_2), and polysilicon layers (if photogates are present). The high absorption of photons in Si implies that carriers are generated very close to the Si-SiO_2 interface. As a result, it is more difficult for these photogenerated carriers to be collected in the depletion region or potential wells, located deeper in the CCD substrate. The presence of SiO_2 also complicate the sensitivity of CCDs in UV. SiO_2 is typically used as an insulating material to serve as the gate dielectric, to provide lateral isolation between adjacent devices, or to function as passivation layers for the device, and it usually lies on top of the semiconductor substrate (as illustrated in Fig. 2.3). This overlying SiO_2 layer does not impose any concern for visible imaging because SiO_2 does not absorb visible photons. However, as the wavelength decreases into the DUV region and below, photon absorption in SiO_2 becomes more significant. This means the incident UV photons are absorbed and intercepted by the top SiO_2 layers, and cannot effectively reach the detector substrate for signal charge generation.[2] Consequently, it is a challenge to manufacture a solid-state detector for UV photons by using conventional design strategies.

CCD manufacturers have adopted a variety of techniques to improve UV performance and to minimize the effect of the UV absorption by the multilayer structure on top of the solid-state imager. These techniques fall into two categories: a direct imaging technique with structural or design modifications to the CCD device, and an indirect imaging technique where the incident UV photons are converted to visible photons by a conversion medium such as phosphor coating. The common UV-enhancement approaches will be highlighted in the next few sections.

3.3.1 Virtual-Phase CCDs and Open-Electrode CCDs

The virtual-phase CCD architecture removes the need for two of the three CCD photogates (in a three-phase clocking scheme) by using suitable implants in the Si to create a series of potential steps that confines the generated charge within the pixel [12]. Figure 3.6 shows a CCD transport section driven by a virtual-phase clocking system. The non-clocking phase, or the virtual phase, is internally biased by the appropriate p-type channel implants that are internally tied to the p-type Si substrate and fixed to the substrate voltage.

---

[2]Refer to Chap. 5 for the optical absorption characteristics of SiO_2.

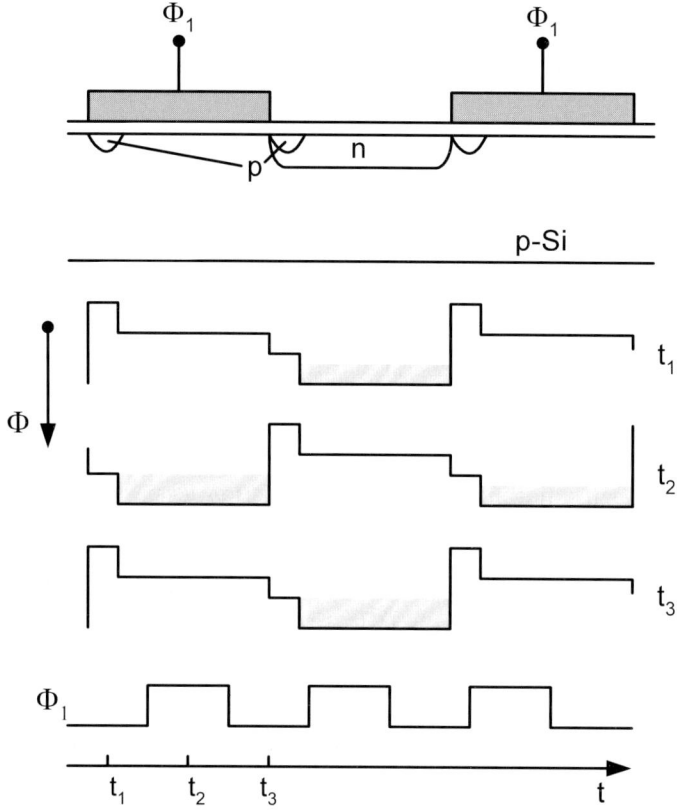

Fig. 3.5. The cross-section of the CCD channel of the open-electrode CCD (adapted from Theuwissen [12])

With this design, the number of gate electrodes, required for the operation of a photogate-based area-array CCD, is considerably reduced. Since the Si surface is only partially covered with the gate material, there is a higher probability that the UV photons will be absorbed in the Si. Therefore, higher QE can be achieved. However, the drawbacks of this configuration is the limited charge-storage capacity compared to that of the standard 2-phase or 3-phase CCD transport system and the high sensitivity to process variations.[3]

Open-electrode CCD, also referred to as open-phase pinned CCD, is based on a similar design approach. The construction of the open-electrode CCD closely resembles that of the virtual-phase CCD, in which one clock phase is defined by implantations (i.e., doping) and is not covered with the gate material. Thus, the absence of the gate material in the open-phase or the virtual-

---

[3]For more information on standard 2-phase or 3-phase CCD transport system, refer to [12] and [14].

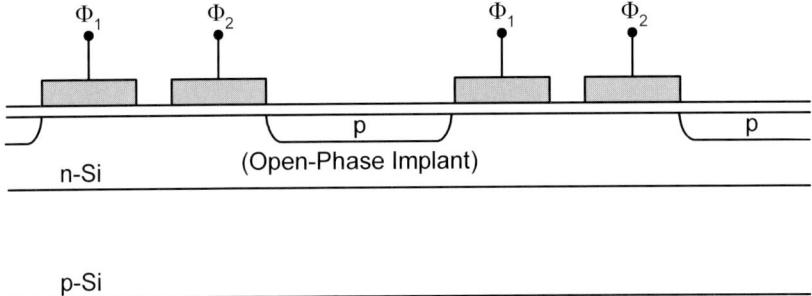

Fig. 3.6. (a) The cross-section of a CCD transport section driven by a virtual-phase clocking system, (b) the corresponding potential well illustration, and (c) the timing diagram of the transport system (adapted from Theuwissen [12])

phase allows more UV photons to penetrate the Si substrate. The principal difference of the two CCD designs is in the definition of the clock phases with polysilicon gates. The virtual-phase CCD supports only one polysilicon gate clock phase, Φ_1, as indicated in Fig. 3.6. In contrast, the open-electrode CCD can support two phases, Φ_1 and Φ_2, as illustrated in the cross-section diagram in Fig. 3.5 [12]. Here, the open-electrode CCD consists of two clock phases (or gate regions) underneath the polysilicon, and a third gate that is defined by the open-phase implantation. The shallow p-doped implant region serves as the pinning layer to equalize the interface potential to the channel stoppers. The deeper n-type implant layer increases the channel potential locally for the signal charge collection. In the open-electrode configuration, the transport of the charge packets is entirely controlled by an external clocking, unlike the virtual-phase configuration where the direction of the charge transport is frozen by the local implants under all the gates. The open-electrode CCDs demonstrate a higher charge transfer efficiency (CTE) and flexibility of clocking than those of the virtual-phase devices. The gate structure can be optimized for a higher sensitivity by choosing the ratio of the open-electrode area to the total CCD cell area to be as high as possible.

3.3.2 Backside-Thinned Back-Illuminated CCDs

To obviate the problems of light absorption (e.g., due to the polysilicon gates) and the reflection on the CCD's front surface, it is possible on target the illumination on the backside of the CCD, where the absorbing gate material is absent. However, to bring the generation sites of the charge carriers as close as possible to their collection sites near the front surface, the Si substrate must be thinned so that only the epitaxial layer remains. Typically, the thickness of a front-illuminated CCD is approximately 500 μm, compared to the thickness of a back-illuminated device which is whittled down to 45 μm or less [34]. The simplified structure of a backside-thinned back-illuminated photogate-based

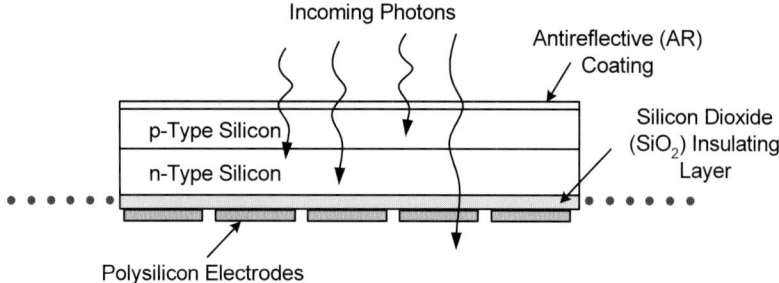

Fig. 3.7. The cross-section of a backside-thinned back-illuminated CCD sensor with photogates arranged in an area array format

CCD is displayed in Fig. 3.7. This design allows photons to directly enter the epitaxial layer of the Si without passing through any gate structure. Furthermore, the design enhances the likelihood for the charge, released near the back of a CCD, to be collected by the potential wells at the front of the device. The net result is a dramatic improvement in the QE, which was illustrated in Fig. 2.6.

A thinned backside-illuminated CCD offers a higher QE than that of a conventional front-illuminated CCD across a wide range of the visible spectrum, and with a more pronounced improvement at the shorter visible wavelengths and at UV wavelengths. The response of the back-illuminated CCD can be further enhanced by the application of an anti-reflective (AR) coating on the thinned backside of the CCD sensor [14]. One drawback of the thinned backside CCDs is the expensive production cost of the substrate thinning which involves a non-standard process and requires delicate handling. In addition, this thinning process reduces the chip yield. Although this detection scheme delivers an enhanced response for blue and UV light, the same does not apply at longer wavelengths. The Si substrate is not thick enough to absorb the longer wavelength photons due to their larger penetration depths. Thus, these thinned CCDs become transparent in near-IR and their response to red light is poor. Another shortcoming with the back-illuminated mode of operation is the possible reduction in the effective resolution of each pixel. Because the photogenerated charges are released near the back surface, the control of the front gate's electric-field over the charges is weaker; thus, the charges may diffuse or disperse to neighboring pixels, causing a reduced resolution.

The etched backside can cause problems with the recombination of the photogenerated charge carriers. The e-h pairs are generated relatively close to the back surface, where the surface is not passivated, and is, consequently, characterized by a disordered crystal structure. The loss of carriers due to recombination is significant in this surface. However, it is possible to suppress the recombination loss by adding an enhancement layer to the backside. Such

3.3 Techniques to Improve UV Sensitivity

a layer can be formed by the implantation of a very shallow p+ layer on the back surface that creates an electric field, forcing the photogenerated electrons toward the potential wells near the front surface, before they are lost by recombination. However, the implantation of the enhancement layer after thinning or etching of the substrate poses a potential technological concern, because further high-temperature processing is not possible with the finished wafers.

Back-illuminated CCDs must be thinned to 10 to 20 μm, which corresponds to the Si epitaxial layer. A natural consequence of such a thin structure, laced with electronic circuitry, is that it wrinkles with a height variation of 30 to 50 μm [34]. This is not desirable for astronomy applications which require a focal plane flatness within fractions of a micron. Furthermore, this membrane is fragile and prone to fracture when subjected to flexing or mechanical stress. To obtain a thinned, flat focal plane array, a rigid support substrate can be attached to the CCD frontside prior to the thinning process [35].

A number of UV-enhancement treatments have been demonstrated for thinned back-illuminated CCDs. An example is the treatment of the back surface using the delta doping technique to form delta-doped CCDs, as illustrated in Fig. 3.8 [35, 36]. Using molecular beam epitaxy, fully-processed backside-thinned CCDs are modified for UV enhancement by growing a thin layer of boron-doped epitaxial Si on the back surface. The term, delta doping, refers to the sharply-spiked dopant profile in the thin epitaxial layer. Because of this ultra-thin layer, the recombination of the UV-generated electrons is almost negligible, and internal QE of 100% has been reported for delta-doped CCDs [36]. Application of an anti-reflective coating can further increase the

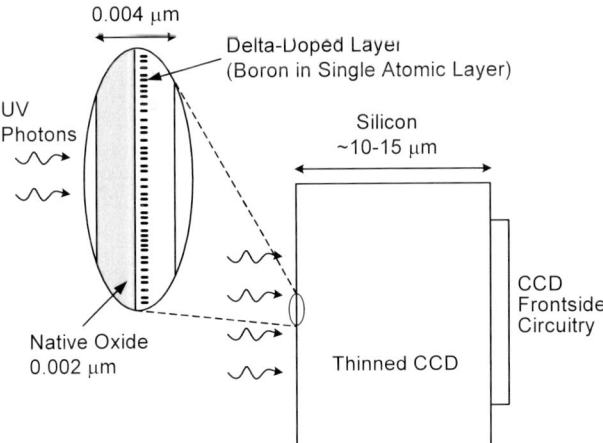

Fig. 3.8. The concept of delta doping for the UV enhancement of backside-thinned back-illuminated CCDs (adapted from [35] and [36])

QE by reducing the reflectivity of the Si at the back surface. In summary, the backside-thinned backside-illuminated CCD design is considered to be very complicated and expensive; yet, it offers the best sensitivity in UV, compared to that of other available CCD implementations.

3.3.3 UV-Phosphor Coated CCDs

To avoid the difficulty of wafer thinning, a less expensive and less complicated approach involves depositing a UV-sensitive phosphor coating on top of the active area of the CCD sensor in a post-packaging procedure. By the appropriate choice of the down-converting phosphor, the UV information is converted to a wavelength that matches the spectral response of the CCD. A typical example of a UV-sensitive phosphor is coronene. When it is excited by UV radiation with wavelengths shorter than 400 nm, coronene fluoresces in the green portion of the visible spectrum, peaking at approximately 500 nm. The sensitivity of a CCD imager that is covered with coronene is extended down to a wavelength of 100 nm [12]. The effect of the phosphor coating on the photoresponse of a virtual-phase CCD is conveyed in Fig. 3.9, where a significant improvement in the QE is attained at UV wavelengths.

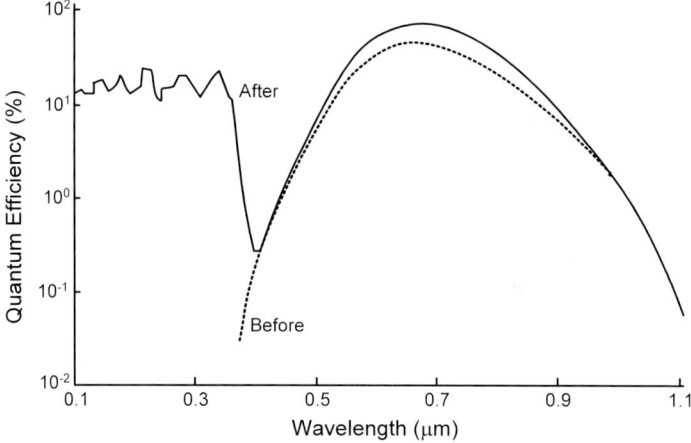

Fig. 3.9. The photoresponse of a virtual-phase CCD before and after being covered with a phosphor coating which converts UV light into green light (adapted from Theuwissen [12])

However, one disadvantage of using this coating technique is a loss in the spatial resolution, caused by light scattering. Due to the isotropic emission of the phosphor layer, only 50% of the visible photons (e.g., 500 nm) are re-emitted in the direction that carries them into the Si depletion layer or substrate, while the other 50% escapes from the CCD sensor. Thus, the

QE of a phosphor-coated CCD at the shorter wavelengths of UV is approximately half of that at 500 nm (the re-emission wavelength). In addition, the optics that are required to re-image the emission of the phosphor coating onto the camera's CCD sensor can introduce image distortion. The limited dynamic range of the usable linear range of the down-converting medium (i.e., the green emission from phosphor) is another limitation of this type of down-conversion systems. Compared to backside-thinned CCDs, the phosphor coating technique is cheaper but it yields a lower QE, and introduces problems such as the photo-degradation of the phosphor coatings and the degradation of the contrast transfer function [12].

3.3.4 Deep-Depletion CCDs

Another approach for improving the UV sensitivity is to use a deep-depletion CCD structure. Here, a very lightly-doped, high-resistivity substrate is used so that the depletion region under the CCD gates is extended to the back of the wafer. Minority carriers, generated by the UV photons which illuminate the rear of the CCD, are swept to the front side by the electric field of the depleted region [12]. This approach to deep-depletion CCDs avoids not only the absorption of the polysilicon gates, but also the thinning that is required for backside-illuminated CCDs with the conventional doping density. A second advantage is that high temperature processing at the rear of the wafer is possible, and a variety of back-side passivating structures can be obtained.

A cross-section of a deep-depletion CCD is shown in Fig. 3.10. It is noteworthy that the back-side implanted (and annealed) p+ layer improves the quality of the back-side of the CCD by decreasing the dark current and increasing the QE of the device. The thickness of this deep-depletion substrate is approximately 150 μm, whereas the resistivity of the substrate is in the range of 4 kΩ·cm to 10 kΩ·cm [12]. A drawback of deep depletion CCDs is that the dark current level increases linearly with the volume of the space-charge region. This is perceivable from the discussion in Sect. 2.3, where it was pointed out that the depletion region and the substrate bulk are two of the key sources for the generation of dark carriers. The presence of the dark current is significant at room temperature, but drops dramatically as the temperature is lowered. Cooling can be accomplished in most scientific UV applications, and therefore, the dark current is not a great concern. For practical applications, deep-depletion CCDs are not as popular as the other UV-enhanced CCD designs, but are favored for X-rays imaging applications.

3.3.5 Other UV Enhancement Techniques

There are other techniques that are available for enhancing the UV responsivity of CCDs, including replacing the standard polysilicon gate material by UV-transmissive indium tin oxide (ITO) gates, and employing a photodiode

40   3 CCD Imaging in the Ultraviolet (UV) Regime

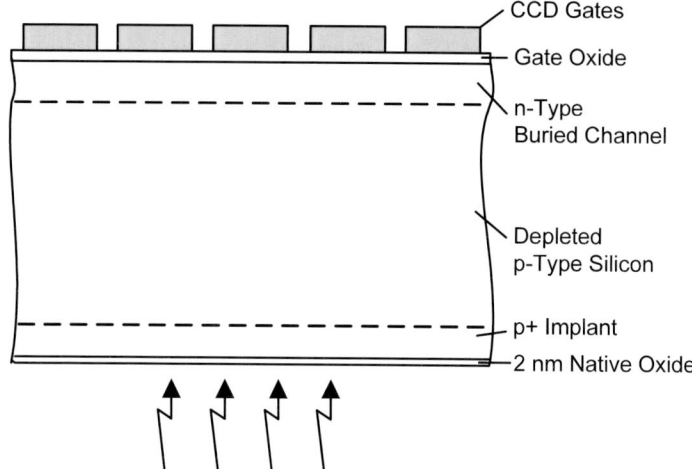

Fig. 3.10. A cross-section of a deep-depletion CCD, built on a high-resistivity silicon substrate (adapted from Theuwissen [12])

instead of a MOS photogate as the pixel sensing element to avoid the use of a polysilicon gate layer. Of the different commercialized devices for increasing the UV responsivity, the thinned back-illuminated CCD sensor exhibits the best device characteristics, but is coincidentally the most expensive.

In this book, a more economical approach for CCD imaging in DUV light is investigated: the frontside-thinned front-illuminated photodiode array CCD. This design is derived from a conventional front-illuminated photodiode-based CCD, but with a thinned overlying oxide layer on top of the imaging area to improve the UV sensitivity. The proposed design has the added advantage that non-standard processing is not required, rendering a more cost-effective design. A detailed discussion of this CCD design and its performance in DUV are presented in Chap. 12.

3.4 Challenges of DUV Detection Using CCDs

Continual advancements in DUV detection have extended the sensitivity of CCD image sensors into the DUV regime. However, there are still several shortcomings of using CCD cameras for imaging DUV excimer laser radiation. First, CCDs are often too sensitive, and require an attenuation of at least eight orders of magnitude [19]. Although attenuation is possible by using optical beam sampling techniques, beam attenuation can introduce other problems in the measurement procedure or in the system. Secondly, the area of the laser beam is sometimes too large to fit on the small active area of the CCD sensor, unless imaging optics are used to resize the beam for measurement. Thirdly, the CCD materials are prone to damage by DUV irradiation, where bleaching (wherein some of the areas of the CCD sensor are reduced in sensitivity) and ablation (where some of the Si material is removed from the CCD array) can occur [19]. These damages can lead to a non-uniform response to DUV radiation across the sensor's surface, a responsivity degradation, long-term instability, and eventually to device failure.

In a test that was conducted at Lambda Physik, the performance of numerous UV detectors (photodiodes and CCDs) were evaluated in long term exposure experiments with 193 nm and 157 nm continuously pulsed laser irradiation. The experiment indicates that most of the UV detectors degrade very fast under the irradiation of 157 nm [37]. The decrease in the spectral response over time ranges from 10% to slightly greater than 50% for Si-based detectors, even when relatively small laser fluences ($<10\,\mu\text{J/cm}^2$) were used [37]. In the literature, it is reported that the fundamental reason for this destruction is the high photon energy, compared to the band-gap energy of the detector material. At the wavelength of 157 nm, the photon energy is 7.9 eV; this is notably higher than the band-gap energy, E_G, of Si at 1.1 eV. Such high photon energy creates a large number of destructive traps to cause the QE degradation [37]. The degradation in CCD performance can also be attributed to the UV-induced defects in the SiO_2 layer and the $Si\text{-}SiO_2$ interface.[4]

In general, the detection of UV light in Si-based CCDs has been a major challenge.[5] This is due to the short absorption depth of UV photons in Si, as well as the strong UV absorption of the frontside layers of the CCD structure (e.g., polysilicon layers, SiO_2 layers, and passivation layers).[6] The short absorption depth signifies that the carriers are generated very close to the $Si\text{-}SiO_2$ interface. This has two implications. First, it is more difficult to collect the photogenerated carriers in the potential well that is situated deeper in the Si substrate. Secondly, there is a higher probability of carrier trapping by the interface states. These factors lead to poor QE and responsivity of CCDs in the UV regime. The strong absorption of UV radiation by the polysilicon gate material, which is present in photogate-based area array CCDs, also limits the CCD performance in UV. Alternative CCD designs that employ a photodiode as the imaging element, or use UV-transmissive gate electrode materials (e.g., indium tin oxide), can avoid the absorption problems related to the polysilicon layers. A more critical concern is that SiO_2 becomes highly absorbing at DUV-VUV wavelengths, and defects can form in the SiO_2 from the radiation. The situation is further complicated by the numerous UV-induced effects and ionizing damages occurring in the SiO_2 layer and the $Si\text{-}SiO_2$ interface, some of which are dose-dependent.

---

[4] A detailed discussion of UV-induced effects in SiO_2 and $Si\text{-}SiO_2$ is presented in Part IV.

[5] The discussion of UV light in this section encompasses all UV wavelengths, including DUV and VUV.

[6] The optical properties of Si and SiO_2 are presented in Chap. 4 and Chap. 5.

Altogether, these factors have a detrimental impact on the responsivity and stability of CCDs in the UV regime.

Although backside-thinned back-illuminated CCDs eliminate the absorption problems related to the frontside polysilicon and SiO_2 structures, the electrical characteristics in a shallow region near the back surface limits the CCD response to short wavelength photons. When the bare Si surface on the backside is exposed to the atmosphere, a thin (<40 Å) native oxide layer develops naturally. This native oxide can interfere with incident DUV photons, and can contain enough trapped positive charge to cause a depletion region to form near the back surface. Photogenerated electrons created in this region drift toward the Si-SiO_2 interface of the back surface, where they become trapped or recombine. This reduces the QE in the short wavelength region. In addition, time- and temperature-dependent changes in CCD response (e.g., QE hysteresis), resulting from the filling and emptying of traps, can occur [10]. Thus, backside-thinned CCDs exhibit a similar degradation and instability in device performance at UV wavelengths as other front-illuminated CCD designs do, due to the ionization damages in the native SiO_2 layer on the back surface.

UV-induced damages in CCDs are examined in the investigation described in this book. Dose-dependent changes in the CCD performance are evident in the experiment at the wavelength of 157 nm. With higher dosages and longer exposures, an overall degradation of the CCD performance is reported. An in-depth discussion and analysis of the UV-induced fluctuations in CCDs are presented in Chap. 12.

The exact causes of DUV-induced CCD degradation have not been completely identified yet. The industry is actively investigating solutions and techniques to address this dilemma. In this book, several mechanisms are proposed regarding the QE fluctuations and the long-term instability, when CCDs are subjected to DUV irradiation. A better understanding of the origins of DUV-induced damages is beneficial for the future development of new-and-improved DUV-sensitive CCD cameras.

Part II

Instabilities in Si, SiO$_2$, and the Si-SiO$_2$ Interface

4 Silicon

Silicon (Si) is an impressive electronic material. Silicon integrated circuits (ICs) and optoelectronics form the basis of the computing and telecommunication industries. Experts in the semiconductor industry agree that Si technology will continue to dominate well into the 21st century [38]. However, the development of microelectronics is unthinkable without the fundamental knowledge of the defects of Si. Depending on the electronic properties, defects can affect semiconductor devices in a desired or undesired manner. In this chapter, the basic optical properties and defects of Si are reviewed.

4.1 Optical Properties of Si

Si is the material of choice for microelectronic fabrication due to its excellent electrical properties. Also, its high optical sensitivity to visible light renders it an ideal material for the fabrication of optical detection devices. The absorption spectrum of Si for wavelengths in the range of 100 nm to 800 nm is displayed in Fig. 4.1. Figure 4.1(b) shows a detailed view of Si's absorption characteristics in the UV region.[1] It is obvious that the absorption depth in Si decreases significantly for UV wavelengths shorter than 400 nm. This is one of the problems in the use of Si-based detectors for UV imaging, where the short absorption depth reduces the carrier collection efficiency in the substrate of the Si detector.

The absorption of photons in the Si results in the generation of electron-hole (e-h) pairs due to the photoelectric effect. The primary function of the CCD is to collect these photogenerated electrons in its potential wells, or pixels, during the CCD's exposure to light. As a result, the photoelectric effect is fundamental to the operation of CCD sensors.

4.1.1 Photoelectric Effect and Si

One of the attributes of Si is its very high sensitivity to light for wavelengths less than 1.1 µm; thus, optical detectors or imagers made of Si are ideal for

---

[1] The absorption characteristics of Si in the visible and in the X-ray to IR spectra were shown in Fig. 2.5 and Fig. 3.2, respectively.

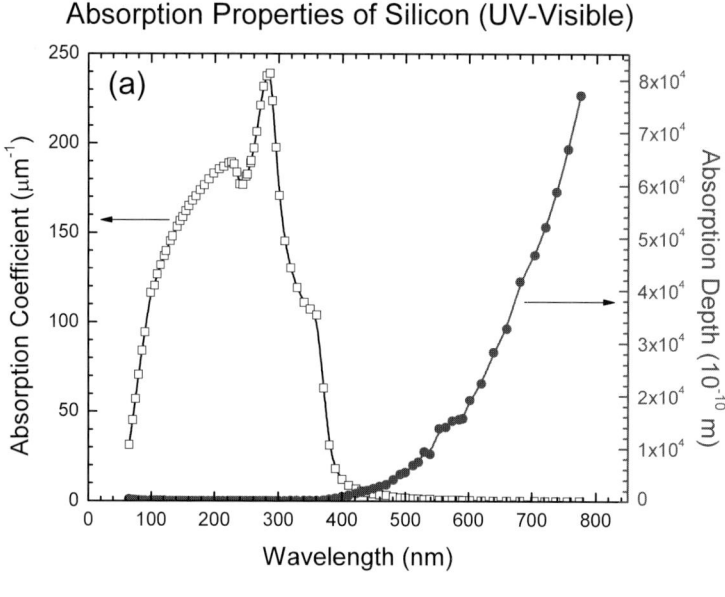
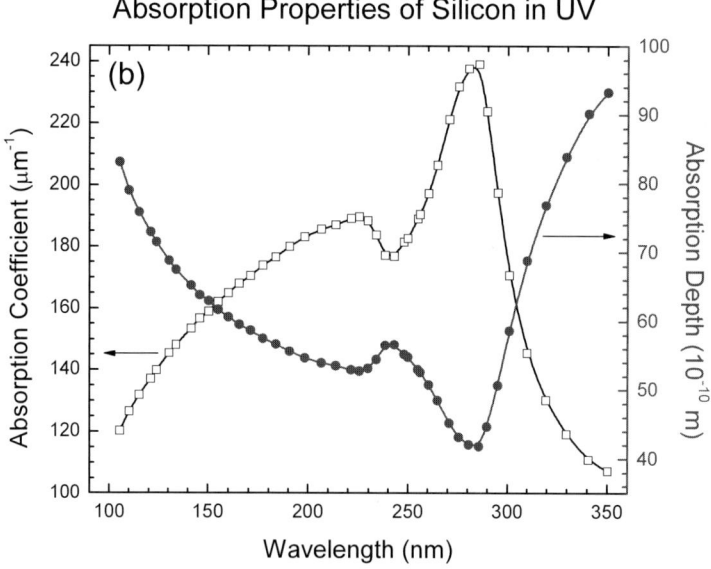

Fig. 4.1. The optical absorption coefficient and absorption depth of silicon in (a) the UV-visible spectrum, and (b) the UV spectrum (data from Palik [39])

visible imaging applications. When the photon energy of the incident light, E_{ph}, is greater than the band-gap energy of Si, E_G (i.e., $E_{\text{ph}} \geq E_G$), the absorption of the photons occurs. The reaction between the Si and the impending light causes the excitation of electrons from the Si valence band to the conduction band. This is known as band-to-band, or fundamental, absorption, which creates one or several, in the case of higher energy photons, e-h pair(s). The e-h pairs become excess carriers and are free to contribute to the conductivity of the material in their respective bands. This is called the *photoelectric effect*. An important aspect of this effect is that a critical wavelength, λ_c, exists in which no electrons can be created for a given material. The critical wavelength is given by

$$\lambda_c [\mu m] = \frac{hc}{E_G} = \frac{1.24}{E_G(\text{eV})} . \tag{4.1}$$

For $\lambda \geq \lambda_c$, the photons have insufficient energy to excite an electron from the valence band to the conduction band. In this case, the photon travels through the material without reacting with the Si lattice. For intrinsic Si with E_G of 1.12 eV, λ_c is 1.1 μm (i.e., the Si is transparent in the far IR) [13]. For extrinsic or doped Si, energy levels exist within the forbidden band-gap, thereby decreasing the effective E_G. As a result, it is possible for photons with wavelengths longer than λ_c to generate a free electron in the Si because the intermediate energy levels can assist in the excitation of valence electrons into the conduction band.

Typically, a photon with an energy of 1.1 eV to 3.1 eV generates a single e-h pair in the Si [14]. More energetic photons with energies greater than 3.1 eV can produce multiple e-h pairs; here, the energetic conduction band electron collides with, and subsequently excites, other valence electrons. For a photon with an energy greater than 10 eV, the average number of electrons generated is

$$\eta_i = \frac{E_{\text{ph}}(\text{eV})}{E_{\text{e-h}}} , \tag{4.2}$$

where η_i is the ideal quantum yield (defined as the number of electrons per interacting photon), and $E_{\text{e-h}}$ is the energy required to generate an e-h pair, which for Si is 3.65 eV/electron at room temperature [14]. Therefore, photons with different energies interact with Si in a slightly different manner, as illustrated in Fig. 4.2. This explains the existence of an absorption spectrum for Si (in Fig. 4.1), and similarly for other materials.

When a photon is absorbed by the detector, the released electron(s) is free to move around in the Si crystal lattice structure. CCDs are designed to store these photogenerated electrons and prevent them from wandering around the lattice [11]. In this manner, a pattern of electrons is collected at the CCD pixels that directly corresponds to the pattern of the incident illumination. The resultant electron charge pattern is transported for readout by using analog electronics, and then digitized by the camera's circuitry

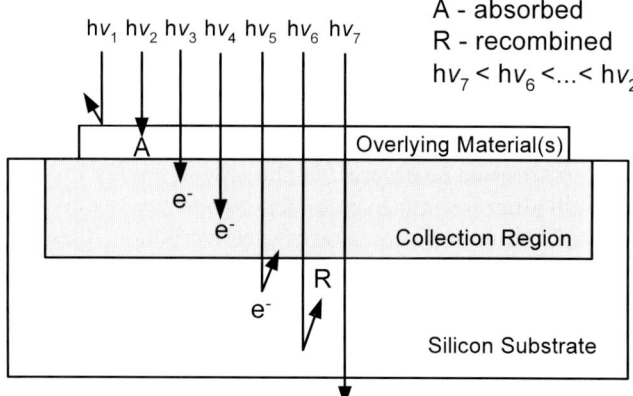

Fig. 4.2. Interaction of photons at different energies with silicon (adapted from [13])

to give an accurate digital representation of the object that is imaged by the CCD sensor.

4.2 Defects in Si

Although single crystalline Si is considered to be the best material in existence for microelectronics [40], defects do exist in the Si lattice and their effects should not be neglected. Possible lattice defects in Si include point defects, dislocations, and grain boundaries.

The lattice defects in microelectronic Si require very strict control. It is possible to avoid larger defects such as grain boundaries, dislocations, and sizeable precipitates in Si crystals. However, there is no way to avoid point defects or small agglomerations of point defects, also known as bulk microdefects (BMDs), crystal originated particles or pits (COPs), or localized light scattering (LLS) defects. Point defects (e.g., vacancies and self-interstitials) are present in thermal equilibrium during the crystal growth, and cannot disappear under most conditions [40]. Consequently, the equilibrium point defects will be either frozen-in during cooling of the Si crystal, or form agglomerates which constitute the bulk microdefects. Usually, these microdefects are few and small; thus, it is not easy to detect them, and they are often below the detection limits of the most advanced methods. Historically, however, Si defects are periodically rediscovered because devices are becoming steadily smaller and more sensitive, and their (consistently negative) influence on device properties continues to emerge [40].

Point defects in Si are of prime importance for three independent reasons. First, point defects are the vehicles for the important diffusion processes of dopants and other foreign atoms. Since there is a definite contribution from the self-interstitials or vacancies that are present in the thermal equilibrium, the diffusion of dopants in Si is much more complicated than in other elemental semiconductors and is still a subject of study. Secondly, point defects are unavoidable because they are present in thermal equilibrium. This limits the perfection of Si single crystals. Also, the agglomeration of point defects during crystal growth introduces microdefects that can lead to more detrimental effects in the device properties. Lastly, fundamental knowledge of the defects in Si is critical for the development of microelectronics. Depending on the electronic properties, defects affect semiconductor devices in a desired or undesired manner.

Point defect equilibria in Si are much more complicated than in other elemental crystals and not very well understood. In particular, Si seems to be the only elemental crystal so far, where self-interstitials are present in thermal equilibrium in concentrations that are comparable to vacancies. (Otherwise, their concentration is consistently much lower.) This implies that both the vacancies and self-interstitials are involved in the formation of BMDs, and that the diffusion of substitutional impurities (e.g., dopants) in Si is likely to be more complicated than in other semiconductors and elemental crystals [41].

Larger defects (e.g., dislocations, grain boundaries and precipitates) are also objects of intense investigations for the following reasons. First, these large defects are present in the polysilicon (poly-Si) that is used for solar cells and MOSFETs, and will influence the technology and the device properties. It is not known, if these defects are electrically-active (i.e., they act as recombination centers), and what determines the level of these activities. Secondly, the larger defects can be formed during the processing of Si. In most cases, their presence inadvertently introduces deadly effects, but sometimes with intentional purposes (e.g., intrinsic gettering). Lastly, larger defects are studied to learn more about defects in covalently bonded crystals (e.g., the precise atomic structure of grain boundaries) [40].

4.3 Si Wafers for CCDs

Typically, CCD sensors are fabricated on Si wafers that range from four to eight inches in diameter with 12-inch wafers in the outlook. An epitaxial Si layer of approximately 10–20 µm in thickness is usually grown on top of a thick, highly-doped Si substrate ($<0.01\,\Omega$-cm resistivity). The substrate thickness for a 4-inch wafer is approximately 500 µm, and increases in thickness as the diameter increases. For n-channel CCDs, the epitaxial layer is pre-doped with boron during growth to yield resistivity in the range of 10–100 Ω-cm. Most CCDs are designed so that the primary CCD functions, including charge generation, charge collection, transfer, and read-out, take place in the epitaxial layer [14].

The thick Si substrate below the epitaxial layer serves several purposes. It supports the epitaxial layer to allow it to be processed, and provides a good electrical ground plate for the device. In addition, the substrate is optically-dead because it is highly doped (approximately 0.01 Ω-cm). This implies that electrons that are photogenerated in the substrate region recombine quickly with the holes that are provided by the dopant. This characteristic is important to achieve a high charge transfer efficiency (CTE) and a good spatial resolution among the pixels in a CCD sensor [14].

The quality of the Si, as well as the presence of impurities or lattice imperfections, has a profound impact on CCD performance. For example, the charge transfer efficiency (CTE) of a CCD is especially vulnerable to the epitaxial quality. This is because CCDs are often required to transfer very small charge packets through several inches of Si without loss, thus a high quality Si epitaxy with minimal density of defects or traps is essential. If traps are present in the charge transfer channel of the epitaxial layer, they can limit the CTE performance of a CCD sensor.

5 Silicon Dioxide

Silicon is unique among semiconductor materials in that its surface can be easily passivated with an oxide layer of SiO_2. The electrical and mechanical properties of the interface between Si and SiO_2, as well as those of the oxide layer itself, are almost ideal. SiO_2 layers are easily grown thermally on Si or deposited on many substrates. They adhere well, they block the diffusion of dopants and many other impurities, they are resistant to most of the chemicals used in Si processing and yet can be patterned and etched with specific chemicals or dry etched with plasmas, they are excellent insulators, and they have stable and reproducible bulk properties [42]. The interface that forms between Si and SiO_2 has very few mechanical or electrical defects and is stable over time. These properties make MOS structures easy to build in Si, and they imply that Si devices of all types are generally reliable and stable under normal operating conditions. The SiO_2 layer serves a number of functions in CCD sensors and MOS devices. These include use as the gate dielectric layer in MOS structures, as an isolation region laterally between adjacent components on the wafer surface, as a stress relief layer (or pad oxide) under silicon nitride (Si_3N_4) layers during local oxidation of silicon (LOCOS) type processes, as an insulator between metal layers in back-end processing, as passivation layer, and as a mask against implantation or other processing steps. As the SiO_2 layers play quite a dominant role in Si-based devices, the properties of SiO_2 and defects in SiO_2 must be considered to gain better insights on how SiO_2 can affect the performance and stability of CCD sensors during DUV irradiation.

5.1 Basic Properties of SiO_2

5.1.1 Structural Properties of SiO_2

Pure silicon dioxide (also identified as silica with the chemical formula, SiO_2) can be described as a continuous network of SiO_4 tetrahedra. These SiO_4 tetrahedra are linked to one another by shared corners, where each oxygen (O) atom forms a bridge between two Si atoms. Such shared atoms are called bridging oxygen atoms. Figure 5.1 denotes the basic unit of SiO_2 and its principal defects. Defects are regarded as deviations from the regular SiO_2

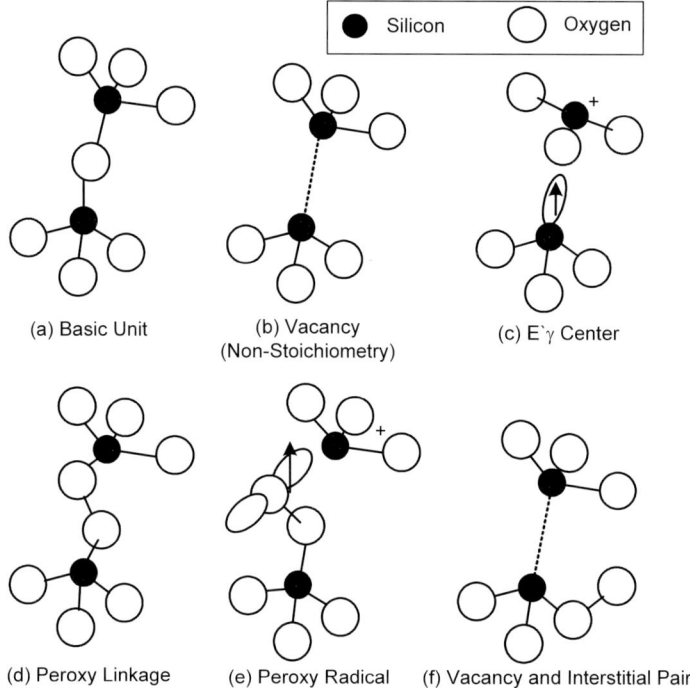

Fig. 5.1. The basic structrual unit of SiO$_2$ and its principal defects (adapted from Nishikawa [43])

structure. They include Si or O vacancies and interstitials, Si-Si or O-O homobonds, and over- and under-coordinated Si or O atoms, as shown in Fig. 5.1(b–f) [43].

SiO$_2$ is classified as crystalline or amorphous. The crystalline SiO$_2$ (c-SiO$_2$) structure exhibits long-range order with an atomic arrangement that is referred to as the *silica lattice* [43]. Various forms of c-SiO$_2$ (also referred to as quartz or crystal) exist, which include α-quartz, β-quartz, tridynmite, and coesite. Contrarily, amorphous SiO$_2$ (a-SiO$_2$) lacks long-range rotational and translational symmetries. However, a-SiO$_2$ (also referred to as glass or fused silica) has short-range order, thus both forms of SiO$_2$ have similar local atomic arrangement. Non-bridging oxygen atoms can be present in a-SiO$_2$, whereas c-SiO$_2$ contains only bridging oxygen bonds. The various crystalline and amorphous forms of SiO$_2$ arise because of the ability of the bridging oxygen bonds to rotate, allowing the position of one tetrahedron to move with respect to its neighbors [42]. The difference between the various forms of SiO$_2$ is characterized by the Si-O-Si bond angle, symbol Θ, which defines the orientation of a given SiO$_4$ tetrahedron with respect to one of its neighbors. The Si-O-Si bond angle for c-SiO$_2$ is limited to a few discrete values; for example, α-quartz is characterized by a bond angle of approximately 144°. In contrast, the Si-O-Si bond angle for a-SiO$_2$ is subjected to site-to-site distribution in the range of 120° to 180°, and the individual tetrahedra are not identical. The bond angles and the interatomic distances of a typical tetrahedron in a-SiO$_2$ are illustrated in Fig. 5.2.

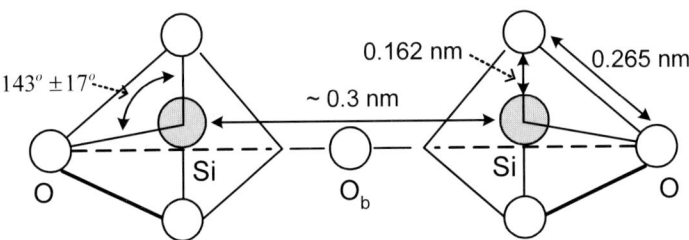

Fig. 5.2. Typical dimensions of the SiO$_4$ tetrahedra in amorphous SiO$_2$ (a-SiO$_2$) (adapted from Barbottin et al. [41])

Oxide layers grown on Si by thermal oxidation are amorphous. If it is grown under well-controlled conditions, thermal oxide displays a high degree of short-range order, and typically yields oxide with a low density of carrier traps. If the processing conditions are poorly controlled, the thermal growth can occur too fast and proceed in a non-equilibrium state. The resultant oxide displays highly disordered areas with strained or dangling atomic bonds, as well as other defects, intrinsic to the SiO$_2$ network. These defects have a significant impact on the transport properties of SiO$_2$. The oxidation temperature also affects the SiO$_2$ film quality. At low processing temperatures, the bulk properties of SiO$_2$ films are not in a complete thermally relaxed state. The films can contain Si-O-H bonds and weak Si-O bonds with a small Si-O-Si bond angle, resulting in an appreciable density of positive charge trapped states, leakage current, and a low breakdown voltage. Although thin oxide films have properties that are different from those of bulk SiO$_2$ glasses, the films share similar types of defects, including oxygen vacancies, oxygen excesses, or other foreign ions at either the substitutional or interstitial positions [41]. Further details on the physical and electrical properties of defects in SiO$_2$ are given in Sect. 5.2.

From the technological point of view, a-SiO$_2$ is considered more important than c-SiO$_2$ because the amorphous phase is readily obtained from the thermal oxidation of Si during microelectronic fabrication. This is the case for MOS devices and CCD image sensors, in which their oxide layers are in a-SiO$_2$. In the following sections, the optical properties and defect characterization of a-SiO$_2$ are considered. For simplicity, the term SiO$_2$ is used and implies a-SiO$_2$, unless specified otherwise.

5.1.2 Optical Properties of SiO_2

Based on optical absorption studies, c-SiO_2 (or quartz) is transparent to wavelengths between 5000 nm and 145 nm [41]. Below 145 nm, the absorption in c-SiO_2 increases abruptly. This absorption can be interpreted as the excitation of electrons from the valence band into the conduction band according to the energy band model. Thus, the band-gap energy, E_G, measured for quartz is between 8.55 eV and 8.4 eV [41]. However, different values of E_G, such as 10 eV, have also been reported for quartz. These deviations are due to the fact that E_G is sensitive to the processing conditions and material quality.

Theory dictates that an effective band-gap exists in any material processing short-range order [41]. Therefore, a band-gap exists in a-SiO_2 which depends on the degree of perfection of the oxide network. The value of E_G decreases as the disorder increases. The experimental E_G values of a-SiO_2 usually range from 8.1 eV to 9 eV, depending on the processing conditions.

Despite the presence of imperfections and defect centers in the amorphous network, a-SiO_2 maintains excellent transmissivity throughout the visible-IR region due to a-SiO_2's wide band-gap characteristic. However, the transparency of a-SiO_2 begins to degrade in the DUV region, where the optical absorption becomes noticeable at wavelengths less than 180 nm. The maximum absorption occurs near the wavelength of 145.9 nm, with an absorption depth of approximately 2.08 nm [39]. The absorption characteristics of a-SiO_2 in the UV region are depicted in Fig. 5.3.

The degraded optical transmission of a-SiO_2 in the UV is evident in Fig. 5.4. A comparison between Fig. 5.4(a) and (b) signifies that the optical loss of a-SiO_2 in the VUV-UV region is four to eight orders of magnitude higher than that in the visible-IR region. Figure 5.4(a) indicates a pronounced absorption at VUV wavelengths that corresponds to the E_G of a-SiO_2 at approximately 9 eV. Thus, the transmission in this region is limited primarily by the band-gap. The presence of defect centers also give rise to absorption bands in the VUV-UV region [43]. In contrast, a-SiO_2 is highly transparent in the visible-IR region for wavelengths from 0.4 to 2.0 μm. However, Fig. 5.4(b) shows that the optical loss of a-SiO_2 increases slightly at longer wavelengths in the near-IR, and is attributed to the vibrational absorption bands in the Si-O network. A few weak absorption bands are present in the visible-IR region, and are related to multi-phonon absorption, Raleigh scattering, UV absorption tail, and structural imperfections and impurities.

5.2 Defects in SiO_2

SiO_2 is found in applications, ranging from common materials such as windows or lens to state-of-the-art technologies such as passivation and gate

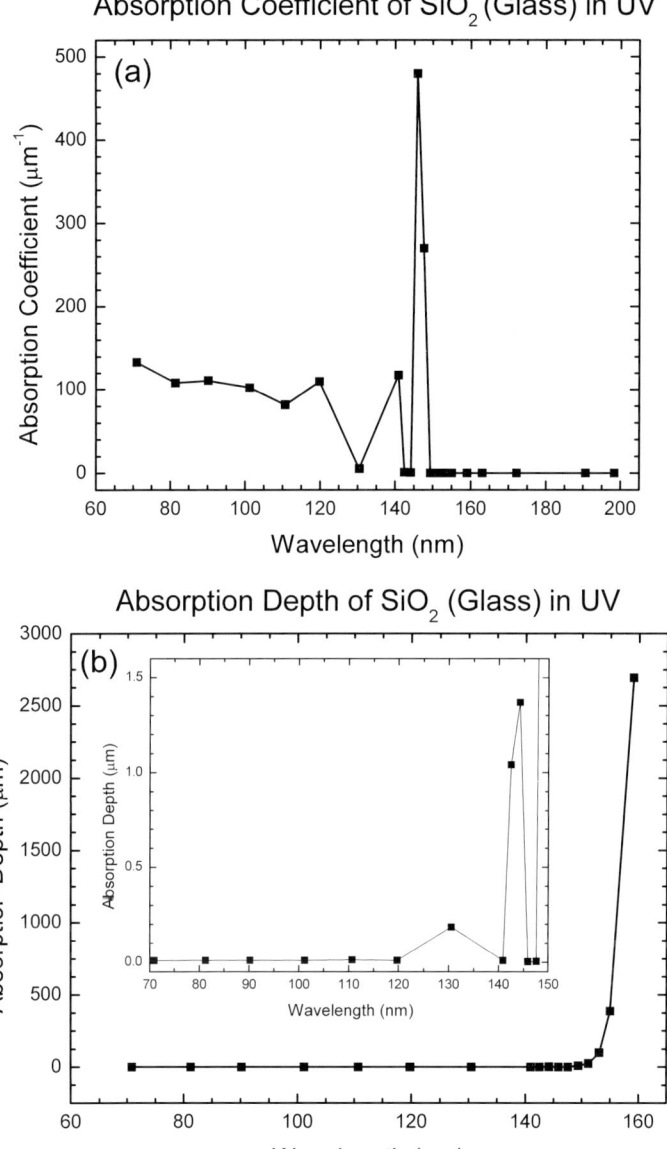

Fig. 5.3. (a) The optical absorption coefficient, and (b) the absorption depth of a-SiO$_2$ at UV wavelengths (data from Palik [39])

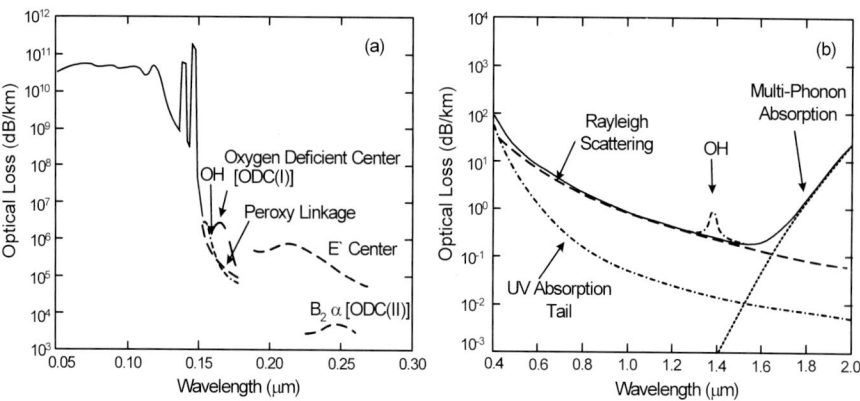

Fig. 5.4. The optical loss of a-SiO$_2$ in (**a**) UV–VUV region, and (**b**) visible–IR region (adapted from Nishikawa [43])

dielectrics of MOS devices, CCDs, and other Si-based microelectronics. Although SiO$_2$ is close to being an ideal insulator for microelectronics, SiO$_2$ is far from being perfect. It contains defects which can have a significant impact on its physical, chemical, structural, optical, and electrical properties. For instance, the presence of electrically-active defects can lead to charging or to charge trapping phenomena that modify the electrical properties of the device, generating instabilities. Optically-active defects can induce phenomena that disturbs the performance of SiO$_2$-based photonics and optical devices. The presence of defects is detrimental to both the production yield and long-term reliability. Therefore, studies of defects are critical in clarifying the microscopic structures and physiochemical aspects of SiO$_2$ [43].

5.2.1 Physical Nature of Defects in SiO$_2$

Defects in SiO$_2$ refer to the structural imperfections and foreign atoms in the SiO$_2$ network. These defects are classified into three categories: microheterogeneities, point defects, and complex defects [41]. Microheterogeneity is defined as slight differences in the structure among essentially identical molecules. In the context of SiO$_2$ defects, microheterogeneities arise when the distribution of Si-O-Si bond angle and that of the Si-O distance become very wide to lead to a change in the amorphous network structure without causing its rupture. However, when the Si-O-Si bond angle and the Si-O distance vary too much, the network can become locally discontinuous. This introduces more foreign atoms, and a wide variety of defects are generated. The second category of defects is known as point defects, which designate any imperfection that disturbs the periodicity of the lattice at one or two atomic sites. A point defect can be classified as *intrinsic* if it is related only to atoms of the original network, such as Si or O vacancies and interstitials,

5.2 Defects in SiO$_2$

Table 5.1. Examples of intrinsic and extrinsic point defects in SiO$_2$ (adapted from [41])

Type of Point Defect	Examples
Intrinsic:	Elongated Si-O bond
	Strained bond
	Broken bond, or dangling bond
	Non-bridging oxygen
	Si or O interstitial
	Si or O vacancy
Extrinsic:	Substitution of a Si atom by a dopant atom (e.g., a group III acceptor atom or a group V donor atom)
	Substitution of an O atom by another anion
	Voids of foreign ions (e.g., Na$^+$, F$^-$, OH$^-$) in the network
	Substitution of a non-bridging oxygen by an univalent anion (or group) (e.g., OH$^-$)

or *extrinsic* if the point defect is related to the presence of a foreign atom. The combination of several point defects leads to the formation of complex defects [41]. Examples of point defects in SiO$_2$ are listed in Table 5.1.

The point defects in SiO$_2$ play a key role in the atomic transport and electronic transport characteristics, because charge diffusion is closely related to the presence of the vacancies. Properties such as photoconductivity, optical absorption, and luminescence are theoretically modeled, based on the defects' electronic structures. In addition, radiation-induced damage in Si-based devices is often associated with intrinsic defects in the Si and SiO$_2$. The various types of intrinsic point defects in SiO$_2$, which will be reviewed next, are divided into three groups: defects related to the Si-O-Si bond, defects related to oxygen, and defects related to Si.

5.2.1.1 Intrinsic Defects Related to the Si-O-Si Bond

Adjacent tetrahedra in the SiO$_2$ network are connected by a common bridging oxygen atom, represented by Si-O-Si. When one of the Si-O bonds becomes strained, distorted or broken, point defects are created. The various abnormal configurations of the Si-O-Si bonds which can cause the generation of intrinsic point defects in SiO$_2$ include an elongated bond, a strained bond, and a broken bond.

The elongation of the Si-O bond is considered to be a defect, since the central O atom moves away from its original position, if the Si-O-Si distance exceeds 3.8 Å. This elongation gives rise to a small dipolar moment [41]. Figure 5.5 represents an elongated Si-O bond. Elongated bonds is a possible source of fixed oxide charges in the bulk of SiO$_2$ or at the Si-SiO$_2$ interface.

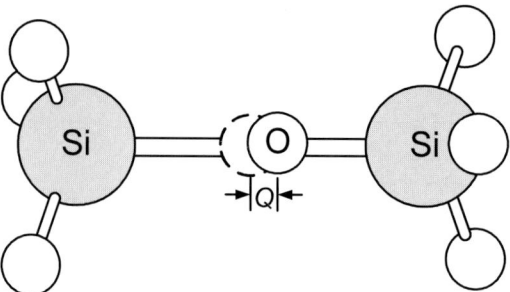

Fig. 5.5. The elongated Si-O bond. The O atom has shifted from its central position (adapted from Barbottin et al. [41])

DUV photons can have sufficient energy to cause the dislocation of an O atom in an elongated Si-O bond, resulting in a fixed oxide charge defect. Subsequently, this can alter the electrical properties and have a large impact on the electrical performance of CCD sensors.[1]

A strained Si-O bond is characterized by perturbed orbitals which differ from the normal configuration, and can be linked to impurity atoms or manufacturing processes that cause modification of the SiO_2 network. Strained bonds can have an impact on the physical, electrical, and optical properties of SiO_2. For example, the different melting behavior between quartz and a-SiO_2 is explained by the concept of the strained molecular bonds in SiO_2. Quartz displays an abrupt melting point, indicating that most of the bonds are broken at the same energy. Conversely, the melting of a-SiO_2 is gradual, indicating that the bonding energies are distributed due to the presence of strained and dangling bonds in the amorphous network [41].

Also, strained Si-O bonds affect the SiO_2 band structure. A distortion or weakening of the Si-O-Si bond causes a shift of the valence band edge, E_V. The strained bonds introduce a continuum of levels near the valence band edge, which disturbs the abruptness of the SiO_2 band edge and alters the band-gap energy, E_G. This is illustrated in Fig. 5.6 and Fig. 5.7(a, b). A shift of E_V by as much as 0.4 eV indicates a variation in the Si-O-Si bond angle distribution, which corresponds to the beginning of an amorphization [41]. This disturbance in the band structure affects the optical absorption of a-SiO_2, especially in the UV region. The VUV absorption edge of SiO_2, which corresponds to E_G, shifts to a lower energy due to the strained bonds. As a result, SiO_2 becomes absorbing at lower photon energies, and thus, the transparency of SiO_2 in the VUV-DUV region is degraded. In addition, strained Si-O-Si bonds play a significant role in the generation of UV-induced defects, which ultimately affect the CCD's optical properties in DUV; these issues will be addressed in Chap. 12.

---

[1] This description is detailed in Chap. 12.

5.2 Defects in SiO$_2$ 59

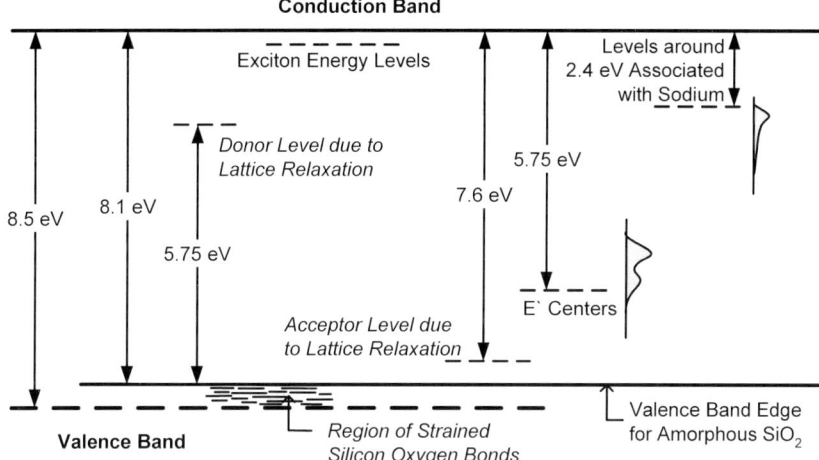

Fig. 5.6. The energy levels introduced in the band-gap of SiO$_2$ due to presence of some defects (adapted from Barbottin et al. [41])

Broken Si-O bonds result from strained bonds which are easily broken by stress (e.g., electric field, temperature, and light radiation). In a broken Si-O bond, the orbitals of the neighboring atoms do not overlap, and the broken bond is eventually accompanied by displaced Si and O atoms. The term *dangling bond* designates a single broken bond at the interface. The amount of energy that is required to break a bond depends on the extent of distortion in the bond. E-h pairs are created in the bond breakage and two situations can develop. In the first case, an e-h pair is generated, where the electron is excited towards the conduction band and the hole towards the valence band to contribute to the normal conduction process. In the second case, an e-h pair is generated, but the electron is immobile in an excitonic state near the conduction band edge, E_C, and the hole is trapped on the broken bond. However, the presence of a coulombic field causes the electron to remain bound to the trapped hole (Fig. 5.7(c)) [41]. Broken bonds can also lead to neutral traps in SiO$_2$. However, these neutral traps disappear when the available bonds are saturated by elements such as H, OH, Cl, O, and F during the annealing procedures. The processes, related to the broken Si-O bond described here, impose fluctuations in the electrical properties of Si-SiO$_2$ devices. In the context of a CCD sensor, the contribution of an electron from a broken bond to the conduction process, or the occupancy of an electron in the mid-gap state associated with a broken Si-O bond, alters the CCD output response. This effect is considered in Chap. 12 to analyze the possible causes of the QE fluctuations that are observed after the CCD sensors are exposed to DUV radiation.

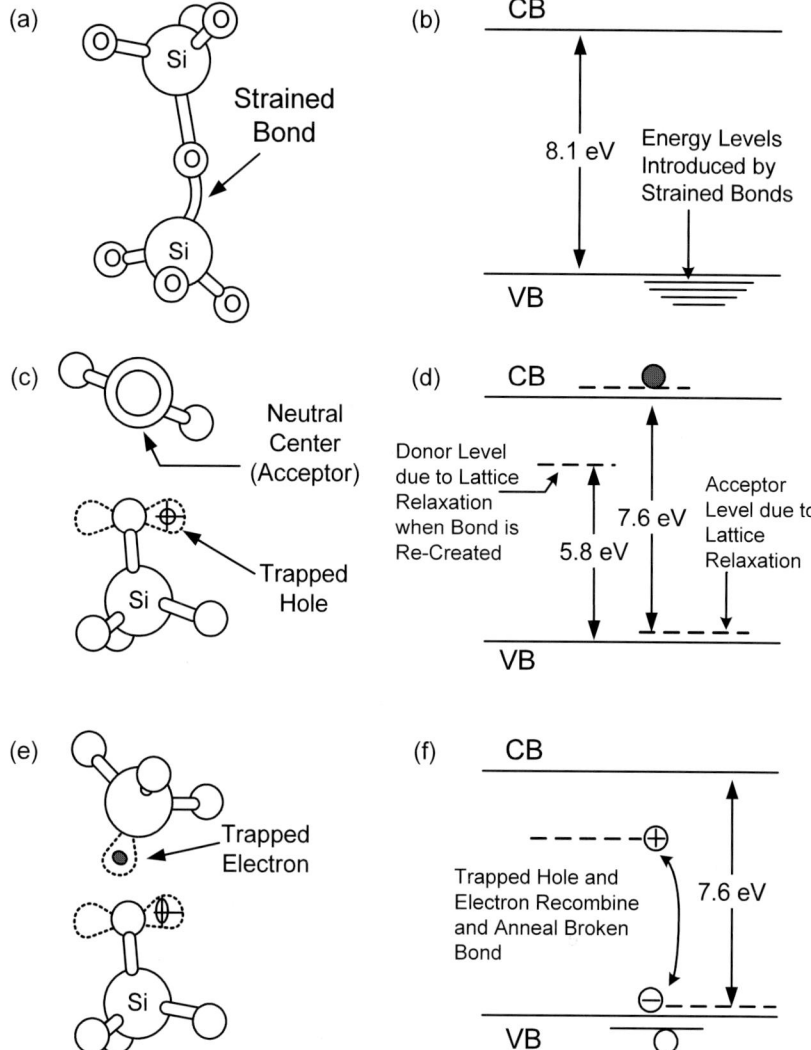

Fig. 5.7. Strained and broken Si-O-Si bonds: (**a**) a strained bond, (**b**) the energy levels introduced by strained bonds, (**c**) a broken bond generating a neutral acceptor center and a trapped hole, (**d**) when the bond is re-created, the network relaxes which gives rise to two new energy levels, and (**e**) the trapping of an injected electron on a Si atom, which (**f**) promotes bond annealing by the recombination with a trapped hole (adapted from Barbottin et al. [41])

A broken Si-O bond can be re-created or repaired by recombining a trapped hole with an electron, where the electron can be supplied from an excitonic state or the conduction band, or it can be a formerly trapped electron. At very low temperatures, the re-creation of the bond is improbable, but for temperatures above 77 K, the bond re-creation process is accelerated by the thermal motion of the ions around their equilibrium position. Bond re-creation leads to a slight relaxation (or re-arrangement) of the surrounding lattice, generating two localized energy levels in the SiO_2 band-gap: a donor level at 5.8 eV above E_V and an acceptor level at 7.6 eV below E_C [41]. This situation is portrayed in Fig. 5.6 and Fig. 5.7. Again, these mechanisms are potential causes of the unstable behavior of a CCD in DUV.

5.2.1.2 Intrinsic Defects Related to Oxygen

In a perfect c-SiO_2 network, an oxygen (O) atom is always bonded to Si atoms, and behaves as a bridge to link two SiO_4 tetrahedra. In the a-SiO_2 network, non-bridging oxygen atoms are present, where some O atoms fail to bridge two tetrahedra and are found in the interstitial positions, or are missing. Non-bridging oxygen is a source of intrinsic point defects. Large concentrations of oxygen-related point defects are often found in thin SiO_2 films. Oxygen-related point defects are generally distinguished into four basic types: non-bridging oxygen (O_n), interstitial oxygen (O_I), oxygen vacancy in place of a non-bridging atom (V_{On}), and oxygen vacancy in place of a bridging atom (V_{Ob}). Each type of oxygen-related point defect is described below.

Non-bridging oxygen (NBO) atoms are quite common in a-SiO_2 films that are not oxygen-deficient. An NBO is represented by $\equiv Si-O^\bullet$, where the notation, $\equiv$, signifies bonds with three separate oxygens and the notation, $\bullet$, signifies an unpaired electron. Since an NBO can readily accept a bond electron, $\triangle$, to complete its outer shell, an NBO constitutes a negatively charged point defect following the reaction which is expressed as

$$\equiv Si-O^\bullet \longrightarrow \equiv Si-O^\bullet_\triangle + h^+ . \tag{5.1}$$

Thus, an NBO appears as a shallow acceptor trap with an energy level, E_T, that is close to E_V. An NBO can also trap holes and thus possesses an amphoteric character. When an NBO captures a hole, it is referred to as an oxygen-hole center (OHC) [41]. The non-bridging oxygen hole center (NBOHC, $\equiv Si-O^\bullet$) is commonly observed in SiO_2 after light irradiation, as detected by Electron Spin Resonance (ESR).

Interstitial oxygen atom, O_I, can originate from two sources: an NBO immobilized in the network as a Si-O group (the most frequent case), and an O atom which is not part of the network. O_I often combines with foreign species (e.g., Na, H) in SiO_2 at room temperature. O_I is an amphoteric center since it can trap holes, as well as electrons. When O_I behaves as an acceptor center, it traps electrons according to the following reaction:

$$O_I^0 \longrightarrow O_I^- + h^+ . \tag{5.2}$$

The trapped electron is weakly bound to such a center in the SiO_2, and the activation energy that is required to free it is much lower than the electronic affinity of the material. When O_I behaves as a donor center, it traps holes by the following reaction:

$$O_I^0 + h^+ \longrightarrow O_I^+ . \tag{5.3}$$

In this case, the energy level, E_T, associated with the O_I donor center is quite deep in the band-gap, since a large amount of energy is usually required to detach an electron from the O_I [41].

Oxygen vacancies, V_O, are categorized into two types. V_{Ob} is related to an oxygen vacancy in place of a bridging atom (i.e., ≡Si• •Si≡), and V_{On} is associated with an oxygen vacancy in place of a non-bridging atom (i.e., ≡Si•). An oxygen vacancy can create a donor trap as follows:

$$V_O^0 \longrightarrow V_O^+ + e^- , \tag{5.4}$$

where an electron is transferred to the Si. At high temperatures (T > 500°C), the oxygen vacancies can migrate under the action of an electric field, towards the Si-SiO_2 interface, where they are neutralized [41]. In this manner, the O vacancy acts a source of fixed oxide charge near the interface.

A commonly-encountered oxygen-related point defect is the E' center, ≡Si•, which is defined as a relaxed oxygen vacancy in SiO_2 and is usually negatively charged. To create an E' center defect, an O atom must be removed from the network, generating two dangling bonds on the Si atoms previously bound to the O atom (Fig. 5.8(a)). Usually, an energy of 7.9 eV is required to break a Si-O-Si bond [41]. One of these dangling Si atoms, labelled as Si_1 in Fig. 5.8(b), traps an electron and remains in a tetrahedric configuration, forming the E' center. Here, the E' center constitutes a trivalent Si atom that has an unpaired electron in a dangling orbital and is back-bonded to three oxygen atoms. The other Si atom, labelled Si_2, relaxes into a planar trigonal configuration. This formation process is depicted in Fig. 5.8. The E' center is perhaps the most commonly observed defect in irradiated thin films of SiO_2 [41]. The E' center is characterized by an optical absorption band of 5.8 eV, thus affecting the optical properties of SiO_2, particularly in the DUV.[2] Also, the E' center can have an electrical influence on Si-SiO_2 devices by acting as a charged oxide center and participating in the hole trapping processes.[3] The DUV-induced effects in SiO_2, related to E' center formation, will be addressed in Chap. 10. E' centers can be generated in SiO_2 by the irradiation of X-rays, γ-rays, electrons, and UV light, or by hole injection, high electric fields, and ion implantation.

---

[2] Refer to Sect. 5.3.1.
[3] Refer to Sect. 5.3.2.

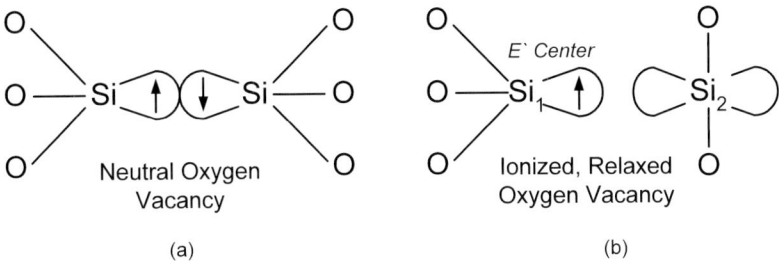

Fig. 5.8. A possible formation process of E' center involves (**a**) removing an O atom from the SiO$_2$ network. (**b**) The defect thus formed is stabilized by an asymmetric relaxation of the two neighboring Si atoms (adapted from Barbottin et al. [41])

5.2.1.3 Intrinsic Defects Related to Silicon

In SiO$_2$, the three main types of intrinsic point defects related to Si include: Si vacancy (V_{Si}), interstitial Si (Si$_I$), and trivalent Si (Si^{3+} or $\equiv$Si^0). Both V_{Si} and Si$_I$ are more difficult to form in SiO$_2$ than in bulk Si due to the fact that the binding energy which links Si^{4+} to O^{2-} in Si-O bond is greater than the binding energy of the Si-Si bond. Typically, a point defect related to Si is formed during the oxidation process. An ionized O atom reacts with a Si atom of the substrate and triggers the displacement of the atom to an interstitial position to form Si$_I$. If the bonds between the two O atoms and Si$_I$ do not form, a mobile Si$_I$ atom is generated [41].

Similarly, the trivalent Si defect (Si^{3+}) is generated during oxidation, or it is formed during high temperature treatment or annealing in hydrogen. This defect plays a role in the formation of some extrinsic defects (e.g., hydrogenated defects) and in some trapping/detrapping activities. A trivalent Si defect behaves as a donor center and is represented as:

$$\text{Si}^{3+} \longrightarrow \text{Si}^{4+} + e^- \quad \text{or}$$
$$\equiv \text{Si}^0 \longrightarrow \equiv \text{Si}^+ + e^- \ . \tag{5.5}$$

Table 5.2 summarizes the key intrinsic point defects in SiO$_2$ [41]. As indicated by the above discussions, some point defects in SiO$_2$ can lead to electrically-active defects, and to carrier traps in the SiO$_2$ network. For example, hole trapping can occur at Si-O bonds in SiO$_2$, or at defects such as an O vacancy, an interstitial Si atom, an E' center, and Si-H and Si-OH groups [41]. Moreover, strained and broken bonds behave like hole traps, and the density of these traps increases with the amplitude of the applied stress. These defects are introduced either by the fabrication process or by high energy radiation (e.g., UV). Therefore, the influence of point defects in SiO$_2$ must be carefully considered, when the UV-induced phenomena in CCDs are examined.

Table 5.2. The point defects of a-SiO$_2$ and their electrical activity (adapted from [41])

Nature of the Point Defect	Symbol or Representation	Electrical Activity
Elongated bond		Generates dipolar moments
Strained bond		Causes a shift of E_V and furthers oxide breakdown
Broken bond		Behaves as a neutral trap
Relaxed re-created Si-O bond		Behaves as an amphoteric carrier trap
Non-bridging oxygen	O_n; $\equiv$Si$-$O$^\bullet$	Amphoteric, but introduces an acceptor level
Interstitial oxygen	O_I	Amphoteric, but introduces a donor level
Oxygen vacancies	V_O; $\equiv$Si$^\bullet$ $^\bullet$Si$\equiv$ or $\equiv$Si$^\bullet$	Introduces two donor levels but the upper one is in the conduction band
E' centers	$\equiv$Si$^\bullet$	A negatively charged, relaxed oxygen vacancy
Silicon vacancy	V_{Si}	Behaves as an amphoteric trap if close to the interface, but also introduces a donor level
Interstitial silicon	Si_I	Introduces two donor levels if near the interface
Trivalent silicon	Si^{3+}; $\equiv$Si^0	Introduces one donor level

5.2.1.4 Self-Trapped Hole

A self-trapped hole (STH), in the form of $\equiv$Si-O$^\bullet$-Si$\equiv$, is considered to be one of the most fundamental point defects in SiO$_2$. STH induces lattice distortions and can be a source of intrinsic interface defects. Radiation can generate STH by trapping a hole at an unperturbed site. Two variants of STH have been identified in SiO$_2$ samples: STH$_1$ and STH$_2$. STH$_1$ is comprised of a hole trapped in the 2p orbital of a bridging oxygen atom in the SiO$_2$ network; STH$_1$ is ascribable to a small polaron (i.e., it is "self trapped" by the relaxation of the surrounding ligands). STH$_2$ is attributed to a hole trapped in two normal (adjacent) oxygens, with the hole rapidly tunneling between the degenerate valence band-edge states on a pair of adjacent oxygens [44]. The proposed models for STH$_1$ and STH$_2$ are schematically presented in Fig. 5.9. STH defects are responsible for the radiation-induced positive charge trapping in MOS oxides, along with E' centers.

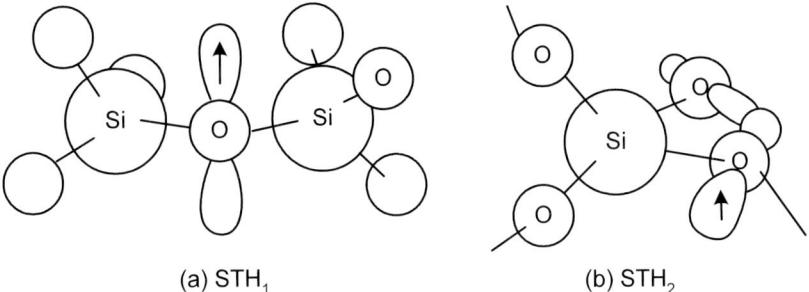

Fig. 5.9. Two types of self-trapped holes in SiO$_2$ (adapted from Nishikawa [43])

5.2.2 Formation Reactions of Defects in SiO$_2$

In this section, the formation reactions for the principal defects in SiO$_2$, of both intrinsic and extrinsic origins, are considered. Table 5.3 lists the principal defects in SiO$_2$ that are induced by the various reactions.

Intrinsic (or native) defects are related to the nonstoichiometry that is observed in the as-prepared materials and the damages introduced by post-fabrication radiation such as X-rays, γ-rays, electrons, and ions. The primary intrinsic defects in SiO$_2$ include oxygen-deficient defects (e.g., E' centers) and excess-type defects (e.g., peroxy radicals/linkages). The E' center, $\equiv$ Si$^\bullet$, constitutes one of the major defects in SiO$_2$ and is generated by the trapping of a hole (or the ejection of an electron), which is represented as

$$\equiv \text{Si} - \text{Si} \equiv \longrightarrow \ \equiv \text{Si}^\bullet + {}^+\text{Si} \equiv + \ e^- \ . \tag{5.6}$$

At higher doses of the deposited energy, the E' center can also be induced by the non-radiative decay of self trapped excitons, where an oxygen atom is pushed outside the normal Si-O-Si network, and is expressed as

$$\equiv \text{Si} - \text{O} - \text{Si} \equiv \longrightarrow \ \equiv \text{Si}^\bullet + {}^+\text{Si} \equiv + \ \text{O}^- \ . \tag{5.7}$$

The displaced interstitial oxygens in this reaction can form O$_2$ molecules and react with the E' center to generate a peroxy radical (PR, $\equiv$ Si $-$ O $-$ O$^\bullet$):

$$\equiv \text{Si}^\bullet + \text{O}_2 \longrightarrow \ \equiv \text{Si} - \text{O} - \text{O}^\bullet \ . \tag{5.8}$$

The same reaction can also take place when the O$_2$ molecules, introduced during manufacturing, react with the E' centers in SiO$_2$. The peroxy linkage (POL, $\equiv$ Si $-$ O $-$ O $-$ Si $\equiv$) is another source of peroxy radical, according to the following reaction:

$$\equiv \text{Si} - \text{O} - \text{O} - \text{Si} \equiv \longrightarrow \ \equiv \text{Si} - \text{O} - \text{O}^\bullet + {}^+\text{Si} \equiv + \ e^- \ . \tag{5.9}$$

The strained Si-O-Si bond is a very important defect to DUV-induced damages in SiO$_2$. With DUV excitation, the strained bond dissociates into

Table 5.3. Common defect formation reactions in a-SiO$_2$ (adapted from [43])

Species	Proposed Reactions	Characteristics
E' center (intrinsic origin)	$\equiv$ Si – O – Si $\equiv$ $\longrightarrow$ $\equiv$ Si$^\bullet$ + $^+$Si $\equiv$ +O$^-$	Electronic excitation by irradiation with neutrons, ions, UV lasers, and γ-rays (>1 Grad)
E' center (extrinsic origin)	$\equiv$ Si – Si $\equiv$ $\longrightarrow$ $\equiv$ Si$^\bullet$ + $^+$Si + e$^-$	Oxygen deficient silica and Si-SiO$_2$ interface
	$\equiv$ Si – H $\longrightarrow$ $\equiv$ Si$^\bullet$ + $^\bullet$H	H introduced during manufacturing
	$\equiv$ Si – Cl $\longrightarrow$ $\equiv$ Si$^\bullet$ + $^\bullet$Cl	Cl introduced from starting materials (e.g., SiCl$_4$)
NBOHC	$\equiv$ Si – OH $\longrightarrow$ $\equiv$ Si – O$^\bullet$ + $^\bullet$H	High-OH type silica
	($\equiv$ Si – O – O – Si $\equiv$) $\longrightarrow$ $\equiv$ Si – O$^\bullet$ + $^\bullet$O – Si $\equiv$	Oxygen excess silica
	$\equiv$ Si + O $\longrightarrow$ $\equiv$ Si – O$^\bullet$	
Peroxy radical	($\equiv$ Si – O – O – Si $\equiv$) $\longrightarrow$ $\equiv$ Si – O – O$^\bullet$ + $^+$Si $\equiv$ + e$^-$	Oxygen excess silica
	$\equiv$ Si$^\bullet$ + O$_2$ $\longrightarrow$ $\equiv$ Si – O – O$^\bullet$	
	$\equiv$ Si$^\bullet$ + 2O $\longrightarrow$ $\equiv$ Si – O – O$^\bullet$	
E'-NBOHC pair	($\equiv$ Si – O – Si $\equiv$)* $\longrightarrow$ $\equiv$ Si$^\bullet$ + $^\bullet$O– Si $\equiv$	Strained Si-O-Si bond
Interstitial O$_2$ ozone	O$_2$ $\to$ 2O or O + O$_2$ $\longrightarrow$ O$_3$	Oxygen excess silica
Chlorine	Cl^0, Cl$_2^-$, Cl + πO $\longrightarrow$ ClO$_\pi$	Silica with high Cl content

an E' center and a non-bridging oxygen hole center (NBOHC, $\equiv$ Si – O$^\bullet$), and is described as follows:

$$(\equiv \text{Si} - \text{O} - \text{Si} \equiv)^* \longrightarrow \equiv \text{Si}^\bullet + {}^\bullet\text{O} - \text{Si} \equiv, \quad (5.10)$$

where * represents a strained state of the Si-O-Si bond [43]. The photolysis of a strained Si-O-Si bond, described in (5.10), is the primary defect formation channel in the SiO$_2$, resulting from DUV excimer laser irradiation.[4]

Extrinsic defects in SiO$_2$ involve impurities such as chlorine (Cl) and hydrogen (H); these impurities are inevitably introduced by the starting materials (e.g., SiCl$_4$) and the preparation processes. Impurities can bond with Si atoms to form Si-Cl, Si-H and Si-OH, which are the potential precursors of E' centers and NBOHCs [43], given by:

---

[4] More details are given in Chap. 10.

$$\equiv \mathrm{Si} - \mathrm{Cl} \longrightarrow \ \equiv \mathrm{Si}^{\bullet} + {}^{\bullet}\mathrm{Cl} \ , \qquad (5.11)$$

$$\equiv \mathrm{Si} - \mathrm{H} \longrightarrow \ \equiv \mathrm{Si}^{\bullet} + {}^{\bullet}\mathrm{H} \ , \qquad (5.12)$$

and

$$\equiv \mathrm{Si} - \mathrm{OH} \longrightarrow \ \equiv \mathrm{Si} - \mathrm{O}^{\bullet} + {}^{\bullet}\mathrm{H} \ . \qquad (5.13)$$

Since a variety of defect formation reactions can lead to the generation of E' centers in SiO_2, it is easy to comprehend why the E' center is a principal defect in SiO_2. Therefore, the presence and influence of E' centers must be carefully analyzed when the defect-related issues and instability of Si-SiO_2 devices are examined.

5.2.3 Behavior of SiO_2 Defects in Si-SiO_2 System

In the fabrication and study of semiconductor devices, the Si-SiO_2 system is of great interest. The original interest was related to the masking capabilities of the SiO_2 against a number of useful impurities. However, the most significant aspect of the Si-SiO_2 system is associated with the properties of the semiconductor-oxide interface (i.e., the Si-SiO_2 interface). The interface properties are influenced by the electronic defects in the SiO_2 layer and in the interface region. Defects at the Si-SiO_2 interface are considered in Chap. 6. In this section, the behavior of the electrically-active defects in SiO_2 are considered.

The presence of imperfections or foreign atoms in the SiO_2 network disturbs the local periodicity of the network and introduces supplementary energy levels in the valence band, in the conduction band, and/or in the band-gap of SiO_2. If the energy level of a defect, denoted as E_T, is located in the forbidden band-gap, the defect is considered to be electrically-active. The defect can behave as a carrier trap, or as an oxide or interface charge, which hinders the electrical performance and stability of Si-based devices.

5.2.3.1 Oxide and Interface Charges

In a Si-SiO_2 structure, the electrically-active defects introduce charges in the oxide and interface layers that alter the electrical properties of the structure. For instance, positive oxide charges can induce a depletion or inversion layer in the underlying p-type Si surface, or an accumulation layer in the underlying n-type Si surface. These electrical activities result in a shift of the capacitance-voltage, C-V, characteristics, and influence other aspects of the Si-SiO_2 structure. The charged defects in the oxide and the interface can be grouped into four types: mobile oxide charges, fixed oxide charges, oxide-trapped charges, and interface-trapped charges [41].

The first type is the mobile oxide charges, Q_m, which are usually due to mobile ionic impurities such as H^+, Na^+, K^+, Li^+, OH^-, and O_2 ions from process contamination [41]. Q_m can drift in an electric field, especially at elevated temperatures, and can contribute to device instability. The second type is the fixed oxide charges, Q_f, which represent the sum of all the immobile charged defects and impurities in the oxide whose state of charge is impervious to bias changes. In practice, whenever SiO_2 is grown thermally on a Si substrate, a thin layer of positive charges are formed and they reside close to the Si-SiO_2 interface ($\lesssim 2$ nm from the surface) [45]. Also, Q_f are generated by irradiation and are important to the UV-induced instability of CCD sensors.

The third type is the oxide-trapped charges, Q_{ot}, which are due to carriers trapped in the defect centers that are located in the bulk of the SiO_2. Q_{ot} are typically associated with e-h pairs that are generated in SiO_2 by ionizing radiation or carrier injection. When the new e-h pairs are separated by an electric field or by diffusion, the carriers can become trapped at defect centers that are initially present, or at traps that are induced by radiation or injection. Electrons are quite mobile in SiO_2, but holes are relatively immobile and are easily trapped [45]. As a result, Q_{ot} are mostly consisted of trapped holes, and are distributed in the bulk of the oxide. The trapped holes constitute a positive charge, which can cause shifts in the device parameters (e.g., flatband voltage shift in MOS structures). Usually, exposure to X-rays, electrons, or other ionizing radiation leads to the generation of Q_{ot}. It is possible to reduce Q_{ot} by annealing the wafers at temperatures in the range of 300°C to 400°C after an exposure to radiation [45].

The last type of charged defect is the interface-trapped charges, Q_{it}, which are due to carriers that are trapped at the Si-SiO_2 interface. Q_{it} can originate from the structural defects and the bond breaking process, and can contain positive or negative charges [41]. The charge per unit area stores in the interface traps is symbolized by Q_{it} while the density of traps per unit area per unit energy in the band-gap is symbolized by D_{it} [45].

Figure 5.10 indicates the location of the four types of oxide and interface charges in a cross section of the Si-SiO_2 structure; the corresponding energy band diagram is also provided. A charged defect in the SiO_2 or the interface sets up a localized electric field that attracts a charge of the opposite polarity in the Si substrate towards the Si-SiO_2 interface. The strength of this localized electric field increases, as the oxide charge approaches the Si-SiO_2 interface. This causes a corresponding change in the attraction force of the oxide charge for the carrier in the Si layer. If the oxide charges are mobile, device instability can result. For example, the threshold voltage shift, ΔV_T, in MOSFETs is induced by mobile oxide charges, Q_m.[5] When Q_m are moved by an electric field in the oxide (from the applied gate voltage), this affects the carriers in the MOSFET channel and causes ΔV_T. Also, the presence of Q_m can generate ion traps, cause ion-induced barrier lowering, and alter the electrical properties of the Si-SiO_2 interface [41]. Therefore, Q_m can be a source of instability in

---

[5]MOSFET is the abbreviation for a metal-oxide-semiconductor field-effect transistor.

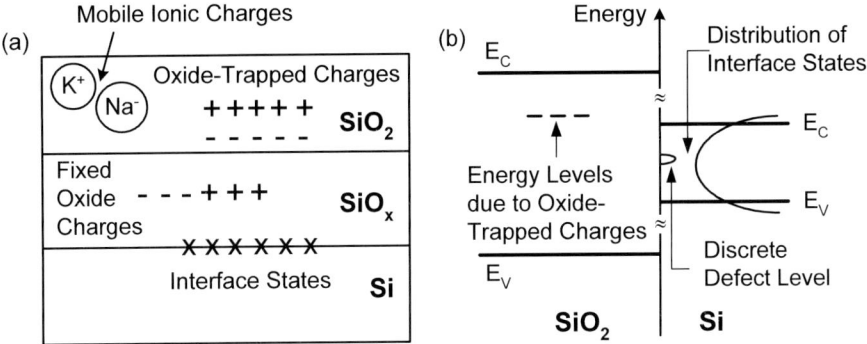

Fig. 5.10. (a) The localization of the four types of oxide/interface charges in a Si-SiO$_2$ structure, and (b) the corresponding energy band diagram (adapted from Barbottin et al. [41])

Si-based devices, and are especially detrimental for devices with an active area juxtaposed to the Si-SiO$_2$ interface, which is the case with MOSFETs and CCDs.[6] The trapping and detrapping behavior of the defects in SiO$_2$ are examined in the next section, accompanied by a discussion on the defects' influence on the electrical stability of Si devices.

5.2.3.2 Classification of Traps in SiO$_2$

A defect in the SiO$_2$ layer or at the Si-SiO$_2$ interface is considered as a trapping center, if it can capture or emit carriers (i.e., electrons or holes). The occupation factor of a trapping center varies according to the stress conditions applied to the device, such as excessive bias, overheating, and an exposure to ionizing radiation. In SiO$_2$, carrier trapping and detrapping are directly related to the carrier injection into SiO$_2$ layer. Carrier trapping gives rise to a space charge (either positive or negative) in the oxide, which modifies the pre-existing electric field and the potential distribution. As a result, carrier trapping can cause device instability such as a breakdown in p-n junctions, a degradation of the current gain in bipolar transistors, a threshold voltage shift in MOSFETs [41]. Consequently, for Si-based devices to maintain an acceptable performance under severe stress conditions, the defect and charge densities in the SiO$_2$ layers must be very low in order to suppress the trapping of carriers. There are various ways to define and classify traps, and the more common classifications are reviewed here.

The first method involves classifying a trap as a *recombination center* or as a *trapping center*. A defect behaves as a recombination center if the discrete energy level, E_T, introduced by the defect, can exchange carriers with both the conduction and valence bands. This implies that E_T is located

---

[6]The instability due to oxide charges are elaborated in Sect. 5.3.2.

near the mid-gap. Secondly, a defect can also behave as a trapping center, if it can capture a carrier and re-emit it into the nearest energy band. This means that the E_T of a trapping center is located near the band edges. When the trapping centers are highly concentrated, they interact. This leads to a broadening of the discrete levels and to the formation of energy bands, called impurity bands, inside the forbidden gap [43]. For example, a trap of level E_{T1}, illustrated in Fig. 5.11, can capture an electron (step 1) and re-emit it later into the conduction band (step 2). Alternatively, the trapped electron can recombine with a hole (step 2a). The former process (trapping and detrapping) is more likely to occur than the latter process, if E_{T1} is far from the mid-gap; in this case, the defect serves as a trapping center rather than a recombination center. This trapping and detrapping process can be repeated with the same traps or with different traps (steps 3, 4, and 5).

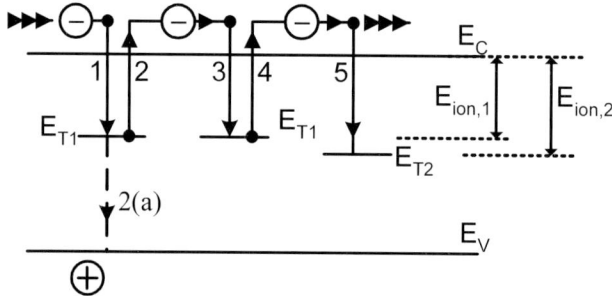

Fig. 5.11. The electron trapping and detrapping by a trap of energy E_T. The capture is represented by steps 1, 3, 5 and the re-emission (detrapping) is represented by steps 2, 4 (adapted from Barbottin et al. [41])

The second approach for defining traps involves categorizing them as *shallow traps* or *deep traps*, depending on their ionization energy. The ionization energy, E_{ion}, of a trap at level E_T, is defined as

$$E_{ion} = E_n = E_C - E_T \quad \text{for electron traps,} \tag{5.14}$$

and

$$E_{ion} = E_p = E_T - E_V \quad \text{for hole traps.} \tag{5.15}$$

The value of E_T, and thus of E_{ion}, is dependent on the environment of the defect and the presence of impurities (e.g., H or Cl) in the vicinity of the defect. By using this concept of E_{ion}, shallow and deep traps can be distinguished as follows. Shallow traps, where a carrier is weakly bound to the defect, possess very small E_{ion} values. This means that the associated defect level, E_T, is close to E_C for an electron trap, or close to E_V for a hole trap. Shallow traps are usually empty at room temperature (300 K), but at low temperatures (e.g. 4 K), the carrier can remain bound to the trap. On the other hand, deep traps possess larger E_{ion} values. In this case, the carrier is tightly bound to the defect, and E_T lies far from E_C (for electron trap) or E_V (for hole trap). Deep traps are often associated with impurity complexes, impurity-vacancy, or impurity-interstitial complexes. Some defects are not shallow nor deep traps, and are described as *intermediate traps*. These traps behave similarly to deep traps [41].

A third way of classifying traps is to consider their state of charge: a *donor trap* or an *acceptor trap*. A donor trap is neutral when it is filled with an electron, and is positively charged when the trap is devoid of an electron. An example of a donor trap is when a Group V element substitutes a Si atom in the SiO_2 network. An acceptor trap is neutral when it is devoid of electron, and is negatively charged when the trap is filled with an electron. Examples of acceptor traps include the substitution of a Si atom in SiO_2 by a Group III element, and the non-oxygen interstitial anions (e.g., OH^-, F^-). Since the same defect can behave as one type of trap for electrons, and as another type for holes, classifying traps, based on donor or acceptor traps, is difficult.

The capture cross section, σ_C, is a more effective way to characterize the trapping behavior of a defect and the ability of a defect to capture a carrier. The concept of the capture cross section can be illustrated by the potential well model. In this model, the presence of a defect leads to a perturbation of the local electrostatic field, and thus, of the potential. This gives rise to a deformation of the conduction (or valence) band around the defect in the shape of a well [41]. E_T represents the energy at the bottom of the well, and σ_C represents the cross section of the well-opening. Depending on the shape of the potential well and on the way it attracts carriers, the range of σ_C values for a trap can be markedly different. As illustrated in Fig. 5.12, three types of traps are distinguished by this model and are described below.

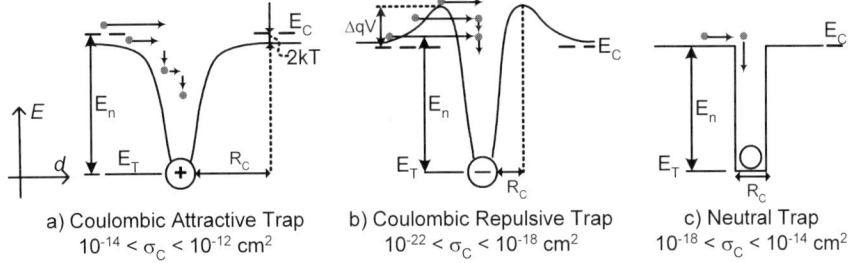

Fig. 5.12. A potential well introduced at the bottom of the conduction band (level E_C) by the presence of a trap (energetically located at level E_T with a capture radius of R_C). The trap can behave as (**a**) a coulombic attractive trap, (**b**) a coulombic repulsive trap, or (**c**) a neutral trap (adapted from Barbottin et al. [41])

First, a trap is *coulombic attractive* if σ_C is large ($10^{-14} < \sigma_C < 10^{-12}\,\text{cm}^2$). Before trapping a carrier, the sign of the charge of a coulombic attractive trap is opposite to that of the carrier. A capture is probable if the carrier approaches the potential well within a distance, r, smaller than the effective capture radius, R_C (Fig. 5.12(a)). After capturing a carrier, the trap center becomes neutral or repulsive, and it cannot capture other carriers of the same type. A coulombic attractive electron trap is usually spatially distributed near the Si-SiO$_2$ interface, and is often related to the presence of sodium ions or other positively charged imperfections that are located in the interfacial region. Also, a trapped hole behaves as a coulombic attractive electron trap. The distribution of this trap in SiO$_2$ depends on the type of stimulation that causes the hole generation (e.g., irradiation, injection) and on the fabrication process of the oxide [41].

Secondly, a trap is *coulombic repulsive* if σ_C is small ($10^{-22} < \sigma_C < 10^{-18}\,\text{cm}^2$). The sign of the apparent charge on a coulombic repulsive trap and that of the carrier it traps are identical, resulting in an electrostatic repulsion between the center and the carrier. Centers of this type are assumed to be surrounded by a potential energy barrier of the height $q\Delta V$ (Fig. 5.12(b)). A carrier is captured, only if it surmounts or tunnels through this energy barrier. Coulombic repulsive traps have complex structures and are often made up of two or more neighboring defects that have opposite polarity or charge.

Lastly, a trap is *neutral* if $10^{-18} < \sigma_C < 10^{-14}\,\text{cm}^2$. In this case, σ_C represents only the cross section of the defect. Neutral trap usually has a very deep potential well (e.g., $E_T = 4\,\text{eV}$). When the well captures a carrier, a large quantity of energy is liberated locally by a multiphonon emission. This triggers not only an electrostatic perturbation in the material, but also a modification of the structure of the defect and its environment. Many defects in insulators possess no apparent charge and mimic neutral trapping centers. Several neutral traps are found in SiO$_2$, especially if oxidation is performed in the presence of water, and/or if the oxide is exposed to a water-laden atmosphere. These traps are located at 4 eV below the E_C of SiO$_2$ and are not accessible optically. However, thermal annealing can annihilate these neutral traps [41]. The trapping mechanism for neutral traps is quite complex. For example, electrons can become trapped in neutral Si-OH groups, where the Si-O becomes negatively charged and a hydrogen is released. Then, the released hydrogen atom either migrates or forms Si-H groups. Also, radiation can perturb bonds and generate neutral traps. Radiation-induced neutral traps usually introduce deep levels in the band-gap, with a spatial distribution that depends on the penetration depth of the irradiated particle.

5.2.3.3 Mechanisms for Carrier Exchange between Traps and Energy Bands

The exchange of carriers between traps and energy bands usually require the presence of an excitation source (e.g., electrical, thermal, and optical). Excitation is applied to the Si-SiO$_2$ structure to inject carriers to fill the traps, or to trigger the detrapping of carriers. Figure 5.13 signifies the possible trapping and detrapping processes in a MOS structure. Similar mechanisms are expected to occur in the Si, SiO$_2$, and interface layers of other Si-based devices such as the CCD sensors. Carrier trapping and detrapping can occur simultaneously in SiO$_2$.

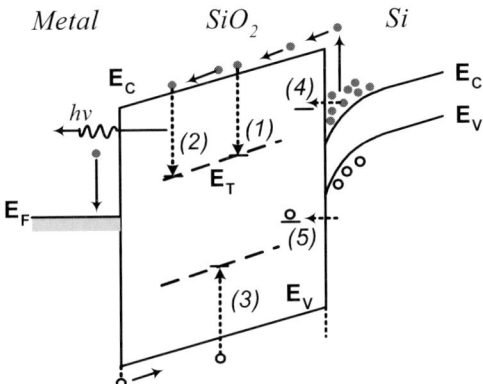

Fig. 5.13. Various trapping mechanisms in a MOS structure. (1) the non-radiative capture of an electron of the SiO$_2$ conduction band, (2) the radiative capture of an electron of the conduction band, (3) the capture of a hole from the SiO$_2$ valence band, (4) the tunnel transition of an electron from the silicon conduction band, and (5) the tunnel transition of a hole from the silicon valence band (adapted from Barbottin et al. [41])

The trapping process begins with the injection of a free electron into the conduction band of SiO$_2$ (possibly by the Si substrate). This electron moves with a mobility ranging from 20 to 40 cm^2 V^{-1} s^{-1}, and can fall into the potential well of a trap in the SiO$_2$. Usually an electron with higher mobility has a smaller probability of being captured. Sometimes, this trapping event is accompanied by a phonon emission (mechanism (1) in Fig. 5.13) or a photon emission (mechanism (2)). Electron trapping can also occur when an electron from the Si conduction band tunnels to an oxide trap located close to the Si-SiO$_2$ interface (mechanism (4)). Likewise, free holes in the valence band of SiO$_2$ can become trapped in an oxide trap (mechanism (3)), which is accompanied by a phonon or photon emission. Also, hole trapping occurs when a hole tunnels from the Si valence band to an oxide trap (mechanism

(5)). Finally, if an oxide trap is located so close to the Si-SiO$_2$ interface that the potential well of the trap spreads into the Si valence band, it is possible to trap holes of only moderate thermal energy (≈ 1 eV). This effect is called *slow hole trapping* [41].

To be re-emitted from a trap (i.e., detrapping), the trapped carrier must gain energy from a stimulation/excitation source. Thermal stimulation can assist in the detrapping of carriers from shallow traps. Optical stimulation can re-emit carriers that are captured by deep traps, provided that the deep trap is accessible optically. Hot carrier stimulation provides a means for detrapping from neutral traps, where the hot carriers transfer part of their energy to a trapped electron [41]. Lastly, electrical stimulation can activate the re-emission of trapped carriers in SiO$_2$ by a sufficiently intense electric field in the oxide, ε_{ox}; however, ε_{ox} must not be too large to cause dielectric breakdown.

Several re-emission mechanisms are possible, depending on the energetic and spatial location of the trap, the intensity of ε_{ox}, and the temperature. For a shallow coulombic trap, with E_T located near the E_C of SiO$_2$, a trapped electron can be re-emitted into the SiO$_2$ conduction band by the action of an electric field via the direct tunnel transition, phonon-assisted tunneling, or Poole-Frenkel effect (i.e., ε_{ox} lowers the energy barrier seen by the trapped carrier, and thus promotes carrier re-emission). For deep traps that are adjacent to the Si-SiO$_2$ interface, the re-emission of electrons into the Si conduction band occurs through tunneling. Such deep traps appear as slow interface states [41]. The detrapping of holes can occur in an analogous manner. Carrier trapping and detrapping mechanisms in SiO$_2$ and in the interface can have a significant impact on the DUV-induced changes in CCDs. For example, the presence of slow interface states at the Si-SiO$_2$ interface can give rise to a time-dependent change in the dark current density during and after the DUV exposure, which are examined in Chap. 12.

5.3 Instability in Si-Based Devices due to Defects in SiO$_2$

The principal defects in SiO$_2$ are classified as optically-active or electrically-active defects. Optically-active defects introduce absorption bands to affect the optical characteristics (e.g., transmission and refraction) of the SiO$_2$ material. Electrically-active defects introduce energy levels in the SiO$_2$ band-gap and participate in the carrier trapping processes. The presence of these defects precipitates instability in Si-based optical and electrical devices. The relevant issues are addressed in this section. Since CCD sensors and MOS devices share very similar structural properties, and since the device instability in MOSFET has been extensively studied by scientists for many years, it is helpful to first review how the presence of defects and charge centers in the oxide or interface affect the stability of MOSFETs. Correlations can then be established between the reliability issues concerning MOSFETs and those of CCDs. Since these reliability issues can also be induced by radiation, the discussion in this section can offer a better perspective of how radiation-induced effects influence the various device parameters of a CCD sensor.

5.3.1 Reliability Issues due to Optically-Active Defects in SiO$_2$

The commonly encountered optically-active defects in SiO$_2$ are summarized in Table 5.4. Some of the defect formation reactions are presented in the Sect. 5.2.2. The presence of optically-active defects has a large impact on the performance of SiO$_2$ components in optical applications such as optical fibers and optics for lithography.

Optically-active defects alter the optical transmission, absorption, or refraction of the SiO$_2$ material. For instance, optically-active defects create color centers that lead to the darkening (i.e., increased absorption) of SiO$_2$-based optical fibers with ionizing radiation, thus degrading the efficiency of

Table 5.4. Optical absorption and photoluminescence (PL) bands associated with optically-active defects and impurities in a-SiO$_2$ (adapted from [43])

Peak Energy of the Absorption Band	Label/Origin	Model	PL Band (and Lifetime)
Defect bands			
7.6 eV	ODC(I)	$\equiv$Si-Si$\equiv$	(a) 4.3–4.4 eV ($\sim$2 ns), (b) 2.7 eV (1–3 ms)
7.6 eV	Peroxy Radical (PR)	$\equiv$Si-O-O$^\bullet$	Not observed
Broad tail (6.5–7.8 eV)	Peroxy Linkage (POL)	$\equiv$Si-O-O-Si$\equiv$	Not observed
6.9 eV	ODC(II)	$\equiv$Si-Si$\equiv$	(a) 4.3–4.4 eV ($\sim$4 ns), (b) 2.7 eV (10.2 ms)
6.7 eV		$\equiv$Si-Si-Si$\equiv$	Not observed
5.8 eV	Si E'	$\equiv$Si$^\bullet$	Not observed
5.15 eV	B$_2\beta$	O-Si-O	4.2 eV, 3.1 eV
5.02 eV	ODC(II) B$_2\alpha$	$\equiv$Si-Si$\equiv$	(a) 4.3–4.4 eV ($\sim$4 ns), (b) 2.7 eV (10.2 ms)
4.8 eV	NBOHC	$\equiv$Si-O$^\bullet$	1.9 eV
	Ozone	O$_3$	1.9 eV
	PR	$\equiv$Si-O-O$^\bullet$	Not observed
3.8 eV	POL (or Cl$_2$)	$\equiv$Si-O-O-Si$\equiv$	Not observed
2.0 eV	NBOHC	$\equiv$Si-O$^\bullet$	1.9 eV (10–20 µs)
Impurity-related bands			
7 eV tail (7.0–8.2 eV)		$\equiv$Si-OH	Not observed
7–8 eV tail	Schumann-Runge continuum	O$_2$	Not observed

the optical fibers for optical communication. Optics (e.g., mirrors and lenses) used in DUV lithography systems are often made of SiO_2, and thus they are susceptible to instability caused by optically-active defects. Synthetic a-SiO_2 is the material of choice for DUV grade optics, since it typically features good transparency, homogeneity, and low susceptibility to laser-induced damage [43]. Nevertheless, synthetic a-SiO_2 is still susceptible to an increased risk of photoinduced damage, when it is exposed to the high power radiation emitted by the intense DUV excimer laser sources in photolithography systems. These photoinduced damages, including color center formation and compaction, usually emerge from the optically-active defects in the SiO_2 network, and can deteriorate the reliability and lifetime of the SiO_2 material.

The radiation-induced color centers introduce absorption bands that degrade the transmission of the SiO_2 optics. Often, the increased absorption results in heating effects, which induce changes in refractive index changes and accelerate optical damage. Of the various color centers, the E' center has the most significant impact on the performance of SiO_2 lens materials. The concentration of E' centers can fluctuate as a function of radiation dose, which deteriorates the stability and lifetime of SiO_2 components. Studies on the DUV excimer laser irradiation of SiO_2 suggested that the E' centers are mainly observed in SiO_2 glasses that are oxygen-deficient, and that the E' centers are generated via the DUV-induced breaking of $\equiv$Si-Si$\equiv$ bonds (refer to (5.6)) [41]. Optically-induced reactions of extrinsic defects, involving impurities such as chlorine and hydrogen, are another source of E' center formations; the associated formation reactions are described by (5.11) and (5.12). Other optically-active defects such as the NBOHC and the peroxy radical can be induced in oxygen-surplus type SiO_2, which follow the reactions that are described by (5.13) and (5.8), respectively.

In addition to the formation of color centers, optically-active defects in SiO_2 can provoke radiation-induced compaction. Compaction usually involves structural rearrangements in SiO_2, driven by the energy that is gained from photon absorption. Compaction affects the reliability of the optical material. These radiation effects or damages in SiO_2 are usually a consequence of the radiation-induced defects, which is understood in terms of radiolytic mechanisms that involve the formation of self-trapped excitons and their non-radiative recombinations [43].

5.3.2 Reliability Issues due to Electrically-Active Defects in SiO_2

Electrically-active defects, in the form of trapping centers or oxide charges, introduce energy levels in the band-gap of SiO_2. They are generated by the high temperature process steps, or by the exposure to energetic particles or radiations (e.g., UV, X-rays, and γ-rays). These defects participate in the trapping and detrapping processes. They also influence the electrical state of a structure by inducing a space-charge layer (e.g., depletion or accumulation of charges) in the Si layer near the Si-SiO_2 interface. Electrically-active defects in the Si-SiO$_2$ system are divided into three categories: oxide traps that have no interaction with the charges in Si, interface-state traps that communicate readily with the charges in Si, and near-interface traps that interact with the charges in Si. The near-interface traps in SiO$_2$ are also identified by names such as slow states, border traps, and switching traps.

A good example of an electrically-active defect in SiO$_2$ is the E' center, which is present as an oxide trap or as a near-interface trap. The E' center is the most commonly observed defect in the irradiated thin film of SiO$_2$, as well as in the bulk SiO$_2$ glasses. A close correlation was reported between the density of the E' centers and that of the hole traps, based on their dependencies on the irradiated doses and processing parameters [43]. This suggests that the E' center is an electrically-active defect, and is involved in electrical processes such as trapping of carriers. In addition, a VUV/UV-induced recycling experiment on undamaged thermal oxides revealed that the charge state of the E' center changes with the UV exposure dose and the irradiation energy [43]. The E' center precursors in SiO$_2$ are activated to a positively charged state of the E' center, when holes are injected into the SiO$_2$ by the exposure to VUV light with a photon energy of 10.2 eV. However, these E' centers are then discharged to a neutral state when electrons are photoinjected from the Si valence band into the SiO$_2$ conduction band by a 5 eV UV excitation. These reactions are perceived as the interconversion between the precursor ($\equiv$Si-Si$\equiv$) and the E' center ($\equiv$Si$^\bullet$ $^+$Si$\equiv$) [43]. Such fluctuation in the charge state of the E' center renders the electrical properties of Si-SiO$_2$ devices to be vulnerable to instability with UV exposure.[7] In addition to the E' centers, impurity-related defects, associated with nitrogen, phosphorus, and boron defects, are other sources of electrically-active defects in SiO$_2$ thin film.

Depending on its location in the SiO$_2$ film, the E' center is also identified as a near-interface trap. The charge state of the E' center that exists very close to the Si-SiO$_2$ interface can be repeatedly switched with an applied bias, and the charge state behaves as an oxide switching traps. An interface-state trap can also originate from the interaction of an atomic hydrogen with an unperturbed SiO$_2$ network. Fluctuations in the charge state of an electrically-active defect is an origin of the instability in Si-based devices [43].

5.3.3 Instability in MOSFETs due to Electrically-Active Defects

Various MOSFET parameters are contingent on the electrically-active defects in the oxide and the interface, including the mobile ions, the bulk oxide traps, and the interfacial charges. These defects can hinder the device performance and stability.

---

[7] Section 10.5 provides a more in-depth discussion on the charging and discharging of the SiO$_2$ layer due to UV-induced fixed oxide charge modifications (such as E' centers).

5.3.3.1 Instability due to Mobile Ions

The flat-band voltage, V_{FB}, and the threshold voltage, V_T, are most sensitive to the mobile ions in the oxide. A redistribution of the mobile ions in the gate oxide affects the V_{FB} of the capacitor-like MOS structure. A shift in the V_{FB}, denoted by ΔV_{FB}, would cause a similar shift in the V_T of a MOSFET (i.e., $\Delta V_T = \Delta V_{FB}$). In turn, a shift in V_T affects other V_T-dependent parameters such as conductance in the linear region, transconductance in the saturation region, and the current-voltage, I_D-V_D, curves [41]. Therefore, the presence of mobile ions in the oxide has an impact on the overall performance of a MOSFET. Similarly, since many of the primary CCD functions (e.g., charge collection, and charge transfer) are strongly dependent on the V_{FB} of the MOS capacitor structure in each CCD cell or pixel, the performance of a CCD sensor is also sensitive to the mobile ions in the oxide. For example, mobile oxide ions can alter the V_{FB}, thereby compromising the charge capacity of a CCD pixel. Moreover, a shift in the V_{FB} of the MOS capacitors in the CCD register affects the charge transfer operation and the charge transfer efficiency (CTE) of the CCD sensor.

5.3.3.2 Instability due to Bulk Oxide Traps

SiO_2 contains defects and impurity atoms in the bulk that act as carrier traps. The bulk oxide traps influence the interface properties differently depending on their distance from the interface. If the bulk oxide traps are located close enough to the interface (e.g., a few nanometers away), they exchange carriers with the substrate in a permanent and spontaneous fashion [41]. In the static regime, this behavior is analogous to that of interface states. Contrarily, if the bulk oxide traps are located far from the interface, they are empty under normal operating conditions. However, the charge contained by the bulk oxide traps can impose an electrostatic influence on the Si substrate. For instance, the charges on the bulk oxide traps can induce a space charge layer in the Si substrate near the interface, which is analogous to the effect of the fixed oxide charges [41]. Therefore, the bulk oxide traps can affect the device performance by altering the electrical property of the SiO_2 layer and the interface.

5.3.3.3 Instability due to Interfacial Charges

Two types of charges can coexist at the Si-SiO_2 interface: the fixed oxide charge, Q_f, and the charge trapped in interface states, Q_{it}. It has been previously discussed that Q_f consists of ionized impurities and charged network defects in the SiO_2. Although Q_f cannot directly exchange carriers with the underlying Si substrate, Q_f can impose an electrostatic field that causes carriers in Si to accumulate close to the interface. This charge accumulation layer affects the interface trapping dynamics and the V_{FB} of a MOSFET or a CCD pixel.

5.3 Instability in Si-Based Devices due to Defects in SiO$_2$

The second type of interfacial charge is related to interface states, which can exchange carriers with the Si substrate [41]. Interface states, typically due to the presence of impurity atoms and lattice defects in the interface, introduce new defect levels inside the forbidden band-gap, and can trap carriers. The number of interface states per unit area, is symbolized by N_{it}, and the quantity of charge trapped in the interface states is symbolized by Q_{it}. Interface states can form discrete levels, or they can be continuously distributed in the energy band-gap. In the latter case, the density of the interface states is symbolized by D_{it}, and is expressed as per unit area and per unit energy. The sign and magnitude of the charge trapped in an interface state depend on the character of the state (acceptor or donor), and on the position of the defect level, E_T, with respect to the Fermi level at the interface, E_{FS}. Typically, not all the interface states are charged; that is $Q_{it} \neq qN_{it}$ [41].

The magnitudes of Q_f and N_{it} are interrelated. They provide a good indication of the quality of the interface: the more the interface is perturbed, the larger the Q_f and N_{it} values. Changes in these parameters imply alterations in the trapping dynamics and in the capacitance of a MOSFET or a CCD pixel. Subsequently, the overall device performance is affected.

Electrically-active defects, including oxide charges and interface states, have an impact on the electrical characteristics of MOS structures. For example, the oxide-trapped charge causes a shift in the capacitance-voltage, C-V, curve of a MOS capacitor, whereas the interface traps tend to stretch/expand the C-V curve. In addition, the presence of interface traps degrades the mobility of carriers in the MOSFET channel, which then alters the transconductance and the gain of the MOSFET [41]. As will be discussed in Chap. 7, radiation can induce electrically-active defects in the SiO$_2$ and in the interface, and thus provoking changes in the performance of a MOSFET or CCD sensor.

6 Si-SiO$_2$ Interface

When a SiO$_2$ film is grown or deposited on a Si substrate, an interface region is formed. Defects and imperfections can exist at the interface and impact the device characteristics. In this chapter, the physical structure of, and the defects at, the Si-SiO$_2$ interface are described. The discussion will provide a rudimentary idea on how interface defects are formed and how they can affect the electrical performance of a CCD sensor.

6.1 Physical Structure of the Si-SiO$_2$ Interface

The interface region can be defined using the *oxygen coordination number*, which is the number of oxygen atoms that is bonded to a given Si atom. For instance, a Si atom in the Si substrate is bonded to four Si atoms and has an oxygen coordination number of zero. The region where the oxygen coordination number varies from zero to four is referred to as the interface region between Si and SiO$_2$ [46].

The atomic structure at the various depths in the thermally-grown SiO$_2$ layer changes as a function of the distance from the Si-SiO$_2$ interface [47]. The oxide is stoichiometric SiO$_2$ to within one monolayer from the the Si substrate surface. A mixed composition of SiO$_2$ species (e.g., Si$_2$O$_3$, SiO, and Si$_2$O) is found closer to the interface. In the intermediate region of the interface, a strained layer, induced by the lattice mismatch between the crystalline Si substrate and the SiO$_2$ layer, is observed and is largely controlled by the conditions of the processing chemistry. Also, a strained region exists in the SiO$_2$ near the interface, and is likely due to a decrease in the Si-O-Si bond angle as the flexible SiO$_2$ network is forced to match with the Si lattice. This is known as the Bond Strain Gradient (BSG) model [47]. As a result, the oxide layer consists of regions with varying composition and atomic arrangement. The bulk of the oxide layer is stoichiometric and is composed primarily of 6-membered rings (with bond angles of 144°) of SiO$_4$ tetrahedra. In contrast, fewer-membered rings (with bond angles of 120°) predominate in the oxide near the interface (<30 Å) [47]. These smaller-membered rings form strained Si-O-Si bonds, which are the principal precursors of the DUV-induced defects in SiO$_2$. In addition to the strained bonds, interface traps and fixed oxide charge centers are present in the interface region. The defects and stresses, generated during fabrication or induced by radiation, can adversely affect the quality of the Si-SiO$_2$ interface.

6.2 Defects at the Interface

The interface normally contains dangling bonds and strongly distorted bonds. These non-idealities are the result of the imperfect matching of the two materials (i.e., Si and SiO$_2$), caused by a deviation in the lattice parameters. Oxygen vacancies in the non-stoichiometric oxide near the interface present another intrinsic source of interface defects [41]. As a consequence, the energy band-gap of the interface is characterized by long band tails for the conduction and valence bands (due to the strained bonds), and a large concentration of dangling bond states that are distributed in the band-gap. The distribution of the interface states in the Si-SiO$_2$ band structure is denoted in Fig. 5.10. Therefore, the band structure of the interface resembles the band-gap of an amorphous material.

Along with the defects in the SiO$_2$, the defects at the Si-SiO$_2$ interface play a crucial role in determining a device's performance. A small density of interface defects is inevitable in all MOS or Si-SiO$_2$ structures. These interface states affect the electrical properties of the Si-SiO$_2$ structures. One of the principal electrically-active interface defects in the Si-SiO$_2$ system is the P_b center. The P_b center is a dangling bond on a Si atom bonded to three Si atoms in the bulk crystal, and will be examined in greater depths in Sect. 6.2.3 and Sect. 6.2.4.

A more general categorization of the Si-SiO$_2$ interface defects classifies defects as *proper* or *improper*. Proper defects are related to the oxidation process or subsequent heat treatments; improper defects constitute all the other defects such as those from alkali ion contamination or radiation damage. The two main types of proper defects in the interface region are the fixed charges and the interface traps. Fixed charges are the charged centers that are confined within 2 nm of the Si-SiO$_2$ interface, and cannot exchange electrons or holes with the lattice. The fixed charges depend on the oxidation and annealing processes, and the surface properties [46]. Interface traps (also referred to as interface states or surface states) are electronic energy levels that are spatially located at the Si-SiO$_2$ interface, and can capture or emit electrons or holes. They are formed by unpaired electrons, and typically arise from the lattice mismatch at the interface, disconnected chemical bonds, or impurities. In the band-gap structure, interface traps are manifested as energy levels in the forbidden band-gap, or in the valence or conduction band. Interface states can be generated by the oxidation and fabrication processes, ionizing radiation, avalanche injection of carriers into the SiO$_2$, and internal photoemission. In most cases, the build-up of radiation-induced interface states exacerbates the noise level and the reliability of the electrical device.

6.2 Defects at the Interface

Depending on their origin, the interface defects can be further categorized as *intrinsic* or *extrinsic*. The fixed charges and interface traps are intrinsic, if they originate from an imperfect Si-O bonding structure along the interface or in the oxide. If they are due to the presence of foreign atoms, they are labelled as extrinsic. Foreign atoms that behave like interface states are not necessarily located at the interface. When foreign atoms are located in the Si layer and close to the interface (within 5 nm), they become indistinguishable from the interface states. These interface states are rapidly charged or discharged by capturing or emitting electrons from the Si bulk. Contrarily, when the foreign atoms are located in the SiO_2 layer and close to the interface (within 2 nm), rapid charging and discharging of the defect states are less likely to occur (except for events that involve tunneling through the oxide such as hole trapping) [46].

Typically, the characteristics of an interface state are influenced by the SiO_2 composition and the surface roughness of the Si substrate. For example, the interface state density strongly depends on the interfacial roughness, whereas the energy level of an interface state is more heavily dependent on the atomic density of the SiO_2 layer [48]. For a SiO_2 layer with a low atomic density (e.g., SiO_2 formed at low temperatures), the interface states are primarily attributable to isolated Si dangling bonds, and the energy distribution of the interface states typically peaks near the mid-gap. Contrarily, for a SiO_2 layer with a high atomic density, there are two distinct peaks in the energy distribution of the interface states, where one peak is above the midgap and the other peak is below the midgap. In this case, the interface states are attributable to Si dangling bonds interacting weakly with a Si or oxygen atom in the SiO_2 layer [48]. The formation reactions, the carrier exchange mechanisms, and the models of interface states are examined in the next sections.

6.2.1 Formation of Interface States

During the thermal growth of SiO_2 on a Si substrate surface, Si atoms with a broken bond at the substrate surface react with the oxidant species to form Si-O bonds. At locations where the O atoms are unable to react with the unbonded Si atoms, H atoms from the oxidation ambient may fill the broken bond, and thus generating Si-H bonds. Normally, Si-H bonds are ideal because they satisfy the broken or dangling Si bonds at the Si-SiO_2 interface by reacting with the H atoms. However, such Si-H bond can get transformed into an interface state, denoted by $Si_3 \equiv Si^{\bullet}$, when it reacts with a radiolytic atomic hydrogen, H^0. This reaction is described by

$$Si_3 \equiv Si - H + H^0 \longrightarrow Si_3 \equiv Si^{\bullet} + H_2 . \tag{6.1}$$

Thus (6.1) is responsible for the formation of interface states. It is possible to passivate an interface state by an exposure to hydrogen, following the reaction:

$$\text{Si}_3 \equiv \text{Si}^\bullet + \text{H}^0 \longrightarrow \text{Si}_3 \equiv \text{Si} - \text{H} \ . \tag{6.2}$$

However, during an exposure to hydrogen, the SiOH group in SiO_2 also reacts to form the NBOHC according to (5.13). This depassivation reaction competes with the passivation reaction at the interface given by (6.2). As a result, the steady-state interface state density is determined by the balance between the depassivation and the passivation reactions [43].

An alternate model for the formation of interface states suggests that the interface traps are generated simultaneously with the fixed charges during oxidation, via the reaction of oxygen with a Si-Si bond at the interface [46],

$$\begin{array}{c}
\vert \\
\text{O} \\
\vert \\
-\text{O} - \text{Si} - \text{O} - + \text{O} \\
\vert \\
\text{Si} \\
\vert\vert\vert
\end{array}
\longrightarrow
\begin{array}{c}
\text{O} \\
\vert \\
-\text{O} - \text{Si}^+ - \text{O} - \\
\uparrow \quad \searrow \\
\text{\textcircled{$\cdot$}} \quad \text{O} \\
\text{Si} \\
\vert\vert\vert
\end{array}
+ \text{e}^- \tag{6.3}$$

The positive Si-O complex,

$$-\text{O} - \underset{\searrow}{\overset{\overset{\displaystyle\text{O}}{\vert}}{\text{Si}^+}} - \text{O} - \\
\text{O} \quad ,$$

is the positive fixed charge, whereas the site,

is the interface trap which contains an unpaired electron [46]. This model assumes that the positive Si-O complex is stable at lower temperatures (<800°C), but the complex is destroyed at higher temperatures (>1000°C) by the following mechanism:

$$-\text{O}-\underset{\underset{\overset{|||}{\text{Si}}}{\overset{\circlearrowright}{\underset{\text{O}}{\nwarrow}}}}{\overset{\overset{|}{\text{O}}}{\text{Si}^+}}-\text{O}- \; + e^- \longrightarrow -\text{O}-\underset{\underset{\overset{|||}{\text{Si}}}{\underset{\overset{|}{\text{O}}}{|}}}{\overset{\overset{|}{\text{O}}}{\underset{\overset{|}{\text{O}}}{\text{Si}}}}-\text{O}- \qquad (6.4)$$

Thus, this reaction simultaneously destroys the fixed charge and the interface trap [46]. According to this model, the interface trap possesses an amphoteric character, as illustrated by

$$\equiv\text{Si}\!\uparrow\downarrow + e^- \rightleftharpoons \text{Si}^-$$
$$\equiv\text{Si}\!\uparrow\downarrow \rightleftharpoons \text{Si}^+ + e^- \; . \qquad (6.5)$$

These reactions signify that an interface state can trap or emit carriers, and thus it forms a generation-recombination center. The carrier exchange dynamics at the interface states will be discussed in the next section.

A common practice for eliminating the interface states at the Si-SiO$_2$ interface is by a heat treatment at 350°C to 450°C in hydrogen [42, 48]. The hydrogen treatment forms Si-H bonds from the Si dangling bonds, as represented by the reaction in (6.2). However, the Si-H bonds are ruptured above approximately 550°C. Thus, heat treatment cannot be performed after the hydrogen treatment, and this places a limitation on the device fabrication process. Moreover, irradiation (e.g., UV and X rays) generates H^0 in the SiO$_2$ layer; and when H^0 diffuses to the Si-SiO$_2$ interface, Si-H bonds are ruptured by the reaction with H^0, resulting in the formation of Si dangling bond interface states. Researchers have been investigating alternative techniques to eliminate interface states and defect states. One group has proposed the use of cyanide treatment, in which the Si wafer is immersed in KCN solutions followed by a rinse in water [48]. This results in the formation of a Si-CN bond from a Si dangling bond at the interface, and leads to a decrease in the interface state density. Theoretical calculations show that the mid-gap state of the interface defect disappears with the formation of a Si-CN bond, and that a CN$^-$ ion is bound to a dangling bond Si atom via the carbon atom. The Si-CN bond energy is calculated to be approximately 4 eV, which is much higher than the Si-H energy of 2.6 eV [48]. Due to the higher bond energy, Si-CN bonds are much more stable than Si-H bonds. Thus, the cyanide treatment can effectively reduce the interface state density in Si-SiO$_2$ structures and improve the electrical characteristics of the Si-based devices [48].

6.2.2 Carrier Exchange at Interface States

Interface states in the semiconductor band-gap play an important role in the determination of electrical characteristics of semiconductor devices such as MOS devices, thin film transistors, solar cells, CCD sensors, and chemical sensors. The operation of an electronic device usually requires the device to be switched between two or more electrical states. The response of the device to an input electrical signal is influenced by the presence of interface states. Device parameters that are affected by the interface states include the transconductance of MOSFETs, and the CTE of CCD sensors. The trapping of carriers at the interface states also contributes to various noise components, including the surface contribution of the leakage currents of p-n junctions and MOS capacitors, and the dark current of CCD sensors. Thus, the electrical characteristics of semiconductor devices are still strongly affected by the interface state density.

Since the interface states can behave as generation-recombination centers, four carrier exchange mechanisms at interface states are possible and are denoted in Fig. 6.1.[1] The first mechanism involves the capture of an electron from the conduction band by an empty defect state at E_T in the interface (mechanism (1)). The second is the emission of an electron into the conduction band by a previously-filled interface state at E_T (mechanism (2)). Similarly, the capture of a hole from the valence band by the filled state, and the emission of a hole by an empty state into the valence band constitute the third and four carrier exchange mechanisms displayed in Fig. 6.1, respectively. Thus, an interface state can trap an electron and/or a hole, depending on the ionization energy of the defect level. The trapping of carriers at interface states influences the operation and performance of Si-based devices.

6.2.3 Models of Interface Defects

Interface states can vary in density, distribution, type, and capture cross section, depending on the processing conditions, crystal orientation, and material properties. Numerous models have been developed to characterize the interface states and are categorized into three groups. The first group considers interface states as inherent to the SiO_2 (e.g., defects in SiO_2 layer); the second group views interface states as a property due only to the Si (e.g., due to Si-SiO_2 misfit dislocation lines); and the last group attributes interface states to the presence of an interface layer or to interface atoms (e.g., dangling bonds) [41]. However, since the experimental characterization of interface states exhibited very diverse outcomes, a comprehensive and unified model that fully describes all the properties of the interface states is still

---

[1]These carrier exchange mechanisms at interface states are analogous to the carrier trapping and detrapping processes in the electrically-active defects in SiO_2, which were described in Sect. 5.2.3.

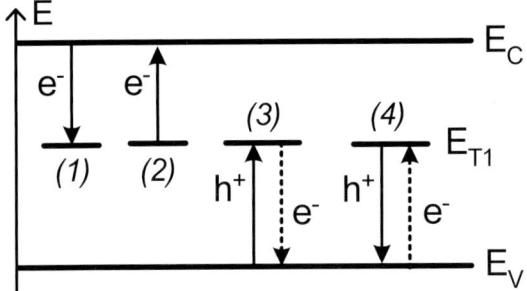

Fig. 6.1. A representation of the four exchange mechanisms at the interface states: (1) electron capture, (2) electron emission, (3) hole capture, and (4) hole emission (adapted from Barbottin et al. [41])

unavailable. The section summarizes some of the popular models for interface states and for fixed interface charges.

6.2.3.1 Model of Interface States based on Defects in SiO_2

One of the interface state model attributes interface states to the deep trap centers in SiO_2 [41]. This model identifies the trap centers in SiO_2 that are located close to the interface to be a source of interface states, and the defects centers that are unaltered by an external signal are a source of fixed oxide charges. Based on this model, a possible candidate for an interface state is the color center, called the B_2 center, in SiO_2. The B_2 band lies at 5.1 eV above the E_V of SiO_2, and is close to the E_C of Si, as shown in Fig. 6.2. Consequently, if the B_2 center is located close to the interface, it is likely to interact with the carriers in the Si conduction band and behave like an interface state.

6.2.3.2 Models of Interface States based on Interface Atoms and Disorder

Models that are based on the assumption that interface states are due to interface atoms have continually evolved, and have facilitated the development of an interface state model that provides good agreement with the experimental results. Some of the more prevalent models in this category include: the disordered interface model, the dangling bond model, the P_b center model, and the trivalent silicon model.

The *disorder interface model* suggests that the disorder at the interface, related to the formation of Si-OH and trivalent Si groups, can result in donor and acceptor centers that act as interface states. This model incorporates the fact that the interface state density, D_{it}, depends on the orientation, oxidation rates, and annealing behavior [41].

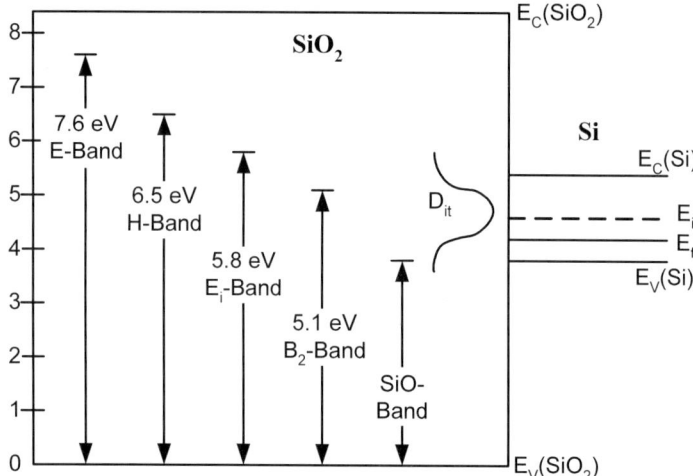

Fig. 6.2. An energy band diagram of the Si-SiO$_2$ structure showing different color centers in irradiated SiO$_2$. The B$_2$ center is a possible source of interface state. D_{it} represents the interface state density (adapted from Barbottin et al. [41])

The *dangling bond model* provides a simple approach for modeling interface states by examining the dangling bonds on either side of the interface. A dangling bond is characterized by an unpaired electron. For a dangling bond located on the Si side of the interface, usually less energy is needed to transfer the unpaired electron of a Si dangling bond up to the Si conduction band than to excite a valence electron to this level. Consequently, this dangling bond makes the interface state to appear as a donor state. However, for a dangling Si atom located at the interface, it tends to complete its shell by accepting an electron; in this situation, this interface state acts as an acceptor state. Therefore, the dangling bond accounts for both the donor-type and acceptor-type interface states [41].

The P_b *center model* refines the notion of the dangling bond by the introduction of a P_b center at the Si-SiO$_2$ interface structure [41]. The family of P_b-centers consists of P_{b0} and P_{b1}. The former is a trivalent Si atom that is bonded to three Si atoms and possesses one dangling bond; the latter is a trivalent Si atom that is bonded to two Si atoms and one oxygen atom, and possesses a dangling bond. Many experimental results support the P_b model by showing a good correlation between the D_{it} at mid-gap and the concentration of P_b centers, even when different oxidation temperatures, annealing conditions, and crystal orientations are examined.

The *trivalent Si model* attempts to provide a unified approach to account for the interface states, fixed interface charges, and radiation-induced charges, by attributing these defects or charges to three forms of trivalent Si atoms: $Si^{\bullet}_S$, $Si^{\bullet}_O$, and Si^{+}_{OS}. These trivalent Si atoms have slightly different chemical

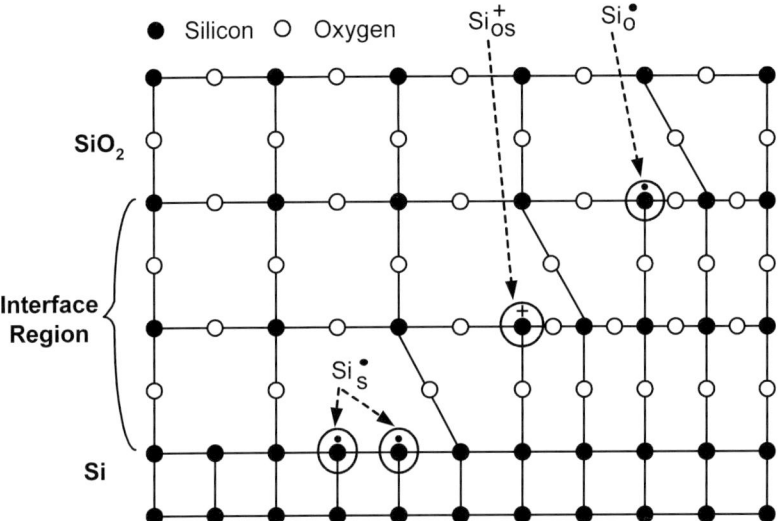

Fig. 6.3. The defect structure of the Si-SiO$_2$ interface according to the trivalent Si model, illustrating the three forms of silicon interface defects (adapted from Barbottin et al. [41])

and electrical properties, as represented in Fig. 6.3 [41]. The first form of Si interface defect is the Si$_S^\bullet$, which is a trivalent Si atom at the Si surface, bonded to three other Si atoms with one remaining dangling bond. This defect is assumed to be an interface trap and is commonly identified as a fast interface state, or the P_b center in other models. The second type is the Si$_O^\bullet$, which is a trivalent Si atom bonded to three oxygen atoms, and is located deeper in the oxide layer. Si$_O^\bullet$ is neutral and is responsible for hole trapping and radiation-induced defects. The third type is the Si$_{OS}^+$, which is an oxide trivalent Si atom, located in the vicinity of the interface. Si$_{OS}^+$ behaves like a very deep hole trap, with a trap energy level that is deeper than the Si$_O^\bullet$ trap. Si$_{OS}^+$ usually remains in a positively charged state.

6.2.3.3 Model of Fixed Interface Charge

The preceding discussion, in Sect. 6.2.2, presented a possible explanation for the formation of fixed interface charge that is deduced from the models of interface states. There is another model of the fixed interface charge, which is based on excess Si in the SiO$_2$ layer and is independent of the interface states. In this model, it is assumed that excess Si ions are accumulated close to the advancing Si-SiO$_2$ interface during oxidation. When the oxidation stops, the atomic distribution in the interface region is suspended, as shown in Fig. 6.4. The excess Si ions near the interface lead to fixed interface charges. This fixed interface charge model explains a number of observations, including a

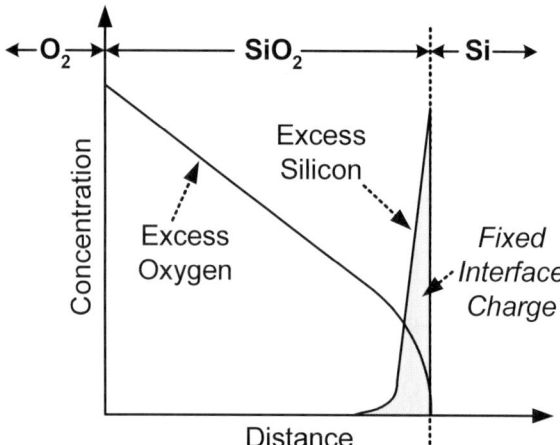

Fig. 6.4. The proposed distribution of excess silicon ions in an oxide film during oxidation to explain the fixed interface charge generation (adapted from Barbottin et al. [41])

low concentration of fixed interface charges at high oxidation temperatures because the excess Si is rapidly consumed, and the annealing of fixed interface charges by the oxygen and Si reaction.

6.2.4 Electrically-Active Defects at the Si-SiO$_2$ Interface

The defects at the Si-SiO$_2$ interface contribute to the electrical instability in Si-based devices. The dominant electrically-active defect at the Si-SiO$_2$ interface is the P_b center. P_b centers are readily generated in thermal oxide, but emerge in slightly different configurations depending on the crystallographic orientation of the Si substrate. On Si(111), the P_b center is a Si atom bonded to three other Si atoms at the interface, and is designated as •Si≡Si$_3$. For the technologically important Si(100), there are two types of P_b centers. One is the P_{b0} center which is essentially similar to the P_b variant on Si(111), but with two possible orientations. The other is called the P_{b1} center which is tentatively assigned to the •Si≡Si$_2$O structure (i.e., a partially oxidized P_b center) [43]. The schematics for the P_b and P_{b0} centers are given in Fig. 6.5. These interface states introduce energy levels in the band-gap of SiO$_2$ and participate in the trapping and detrapping of carriers. Therefore, the presence of these electrically-active interface defects can alter the electrical performance of MOS and CCD structures. For example, interface state generation is a key contributor to the dark current of CCD sensors.

As indicated by the discussion in Part II of this book, a number of defects in the SiO$_2$ and at the Si-SiO$_2$ interface are sources of radiation-induced defects. For example, radiation can generate defects such as the E' center, the NBOHC and the self-trapped hole in SiO$_2$, and the trivalent Si defects

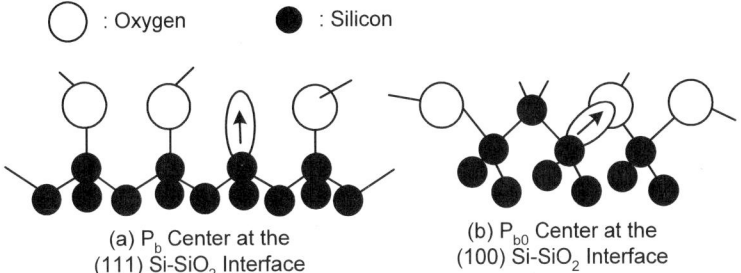

Fig. 6.5. The P_b and P_{b0} centers for the Si-SiO$_2$ interface on Si(111) and Si(100), respectively (adapted from Nishikawa [43])

at the interface, which can subsequently impact the electrical and optical properties of a Si-SiO$_2$ structure. These defects must be carefully considered when the radiation-induced damage in Si-based devices is examined. More in-depth discussion on the radiation-induced defects is provided in Chap. 7 to Chap. 13.

Part III

Effects of Radiation on the Si-SiO$_2$ System

7 General Effects of Radiation

The radiation-induced degradation of device performance is a severe limitation to the application of electronic components in a radiation environment. The capability of semiconductor devices to withstand a specific radiation stress is an unconditional requirement for a variety of space and industrial applications. The sensitivity of a particular device depends on the type of radiation and whether the dose is delivered in a transient burst or over a relatively long period of time [41]. Radiation damage in semiconductor devices is described as a two step process. The primary step is the interaction between radiation and matter, which is explained by the physical phenomena of absorption and scattering effects. The secondary step encompasses the reordering of the distorted regions of the lattice, the charge trapping in the insulators, and the creation of new defects.

Part III focuses on the causes and effects of radiation damage in Si-SiO$_2$ devices. Specifically, the radiation effects on MOS devices and CCD sensors are considered.

7.1 Overview of Radiation Effects on Matter

The permanent damage that is inflicted on a given material by radiation is characterized by two key components: displacement damage and ionization damage. If there is an interaction between the energetic particles from the radiation with the electrons in the material, ionization occurs (i.e., the electrons are excited from the valence band to the conduction band of the material). If there is an elastic interaction between the energetic particles and the nuclei, energy is transmitted to these nuclei and results in atomic displacements [41]. Both types of interaction can generate defects. Defects in the form of atomic displacements occur directly, when the energy that is transmitted to the nuclei is large enough. Defects can also be produced indirectly, when ionization damage occurs in the insulating materials (e.g., SiO$_2$). For semiconducting materials, the primary radiation-generated defects include vacancies and interstitials, and they often interact to form more complex defects.

This section begins with a review of the fundamentals of radiation-and-matter interactions. Since the objective is to study the effects of DUV radiation on CCD sensors, the discussion will emphasize the interaction between electromagnetic radiation and matter.

7.1.1 Displacement Damage

The displacement of atoms from their original positions leads to the formation of electrically-active defects. These defects can be point defects such as interstitials, vacancies, Frenkel defects, or defect clusters [41]. Typically, significant displacement damage is caused by neutrons from a nuclear reactor or from the irradiation of heavy particles. In order to produce atomic displacement (i.e., to remove an atom from its substitutional position and place it in an interstitial site), the transmitted energy, T, from the incident particle needs to be larger than the minimum value, T_d, called the threshold energy. This threshold energy is the approximate energy that is required to break the bonds that bind the displacement atoms to their neighbors (e.g., four bonds in the case of Si). T_d for Si is approximately 25 eV; this energy corresponds to the photon energy from EUV or X-ray radiation. Since DUV photons have an energy less than 10 eV, DUV radiation is unlikely to cause atomic displacement in Si [41].

The defects induced from the displacement damage can affect the electronic properties of matter in three different ways. First, the defects affect the excess carrier lifetime, if they act as recombination centers, and if the defect density induced by displacement damage is comparable to or larger than the original defect density in the material. For example, displacement damage can cause a reduction in the minority carrier lifetime in the Si substrate; this, in turn, has an adverse effect on the gain of bipolar devices. Secondly, the defects induced by displacement damage affect the density of free carriers, if the defects behave as trapping centers with a shallow energy level in the band-gap, and if the defect density is comparable to the doping density. This effect is called carrier removal. Lastly, the defects affect the carrier mobility if they are charged, and if they serve as scattering sites [41].

7.1.2 Ionization Damage

Ionizing radiation breaks atomic bonds and creates electron-hole (e-h) pairs. The radiation is in the form of photons whose energies are greater than the band-gap energy, E_G, of the irradiated material, or in the form of energetic particles such as electrons, protons or atomic ions [49]. Ionization is the process of converting neutral atoms or molecules into ions by removing (or adding) an electron from the atom. The energy required to free an electron from the lattice atom is called the ionization energy, E_{ion}, or ionization potential. Since this process is accompanied by heat from phonon excitation, E_{ion} is larger than E_G. In most cases, E_{ion} is two to three times larger than E_G [49]. The typical values for the ionization energy of Si and SiO_2 are listed in Table 7.1.

Table 7.1. Ionization energy and band-gap energy of Si and SiO$_2$ [49]

	E_{ion}	E_{G}
Si	$3.70 \pm 1\,\text{eV}$	$1.11\,\text{eV}$
SiO$_2$	$18\,\text{eV}$	$8.9\text{–}9.2\,\text{eV}$

The common defects that are created by ionizing radiation include trapped charges, lattice defects, and Frenkel defects [49]. When free carriers are generated by ionization, they either become trapped in the defects that are already present in the material (i.e., trapped charges are formed), or are self-trapped. Self-trapped carriers induce lattice distortions and must be regarded as defects.[1] The creation of e-h pairs from ionization can, in turn, trigger the creation of Frenkel pairs (i.e., vacancy-interstitial pairs). This occurs when the energy transmitted from the incident radiation to the material is converted into a form that is capable of displacing atoms, thus Frenkel defects are generated [49].

In semiconductors, ionization is a transient phenomenon, since the creation of e-h pairs ceases when the irradiation ceases [49]. During irradiation, a stationary equilibrium between the generation and the recombination processes is established within the semiconductor. If an electric field is present, an additional photocurrent can be measured and is proportional to the dose rate. At very high dose rates, the semiconductor device becomes an excellent conductor, since large quantities of e-h pairs are generated by the irradiation. However, burn-out and latch-up effects occur if the current flow is not properly controlled or limited.

In insulators, ionizing radiation first causes the generation of an e-h pair. The photogenerated holes can become trapped to result in ionization damage in the insulator [49]. Hole trapping is due to various reasons such as the large difference between the electron mobility and the hole mobility in the insulator, the differences in the trapping sites, and the differences in the potential barriers for both the carriers at the semiconductor-insulator interface. Hole trapping gives rise to trapped charges within the insulator and to interfacial charges or traps at the interface. Ionization damage in the form of charge accumulation in the insulator is technically termed as the total dose effect [49]. This charge accumulation in the insulator layer imposes fluctuations in the device performance, and is one of the most important damage mechanisms in MOS technology.

Since the number of generated e-h pairs is directly proportional to the amount of energy absorbed by the device material, the total damage is roughly proportional to the total dose of the radiation received by the device. As the total dose increases, the amount of the oxide-trapped charges and the number of interface traps increase monotonically. The radiation hardness of

---

[1] An example is the self-trapped holes, as discussed in Sect. 5.2.

a device is determined by the rate at which the oxide-trapped charges and interface traps accumulate as the total radiation dose increases. For radiation-hard devices, the number of radiation-generated holes, trapped within the oxide layer, is typically less than a few percent of the number created within the oxide layer; the number of radiation-induced interface traps is also less than a few percent of the number of e-h pairs generated within the oxide layer [49]. For radiation-soft devices, the number of trapped holes is usually greater than 50% of the total number of e-h pairs generated; the percentage of radiation-induced interface traps is also larger, although not as much as the radiation-induced trapped holes [49]. Therefore, radiation-soft devices are more sensitive to radiation damage.

7.1.3 Interactions of Photons with Matter

Photons can interact with matter to cause atomic displacement and ionization, and the manner by which this happens is principally determined by their energy,

$$E_{\text{ph}} = h\upsilon ,\tag{7.1}$$

and their momentum,

$$P = h/\lambda ,\tag{7.2}$$

where h is the Planck's constant ($h = 6.62 \times 10^{-34}$ J-s), λ is the wavelength (in cm), and υ is the frequency (in s^{-1}) [41]. Depending on the photon energy, photons can cause ionization in a material by one of the three mechanisms: the photoelectric effect, Compton scattering, and pair production. These three types of interactions are represented in Fig. 7.1.

The photoelectric effect is the phenomenon in which charge particles are released from a material when it absorbs radiant energy. This effect occurs when an incident photon with energy $h\upsilon$ is completely absorbed by the material; the incident photon then interacts and transfers its energy to an orbital electron, ejecting an electron from the atom. The kinetic energy of the released electron is $h\upsilon - E_B$, where E_B represents the energy binding

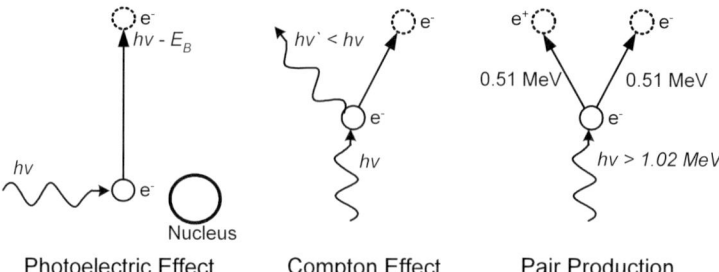

Fig. 7.1. The three mechanisms by which photons interact with matter (adapted from Barbottin et al. [41])

the ejected electron to the atom. Low-energy photons (<0.1 MeV) interact through the photoelectric effect to generate e-h pairs [41]. The probability of a photoelectric interaction decreases, if the photon energy increases, or the atomic number decreases.

The second type of interaction is Compton scattering. This results from the collision of a photon with an electron, and is the principal absorption mechanism for photons with energies between 100 keV and 10 MeV (e.g., γ-rays from nuclear explosion). Here, the incident photon energy is greater than the binding energy, E_B, of the valence electron. Thus, the energy from the interacting photon causes the ejection of an electron from an atom, accompanied by the creation a new photon with a lower energy, $h\upsilon'$; this new photon can then interact photoelectrically with the material [41]. As a result, the photon energy is partially scattered and partially absorbed. The probability of Compton scattering decreases as the photon energy increases, and is relatively independent of the atomic number of the absorbing material.

The third type of interaction is pair production, and involves incident photon with energy exceeding 1.02 MeV [41]. Pair production results in the annihilation of an incident photon and the generation of an electron-positron pair. A positron is a positively charged electron, and has a very short lifetime. The total energy of the electron-positron pair is equal to $h\upsilon$.

For DUV photons, the photoelectric effect and ionization damage are the dominant radiation effects. Defects that are induced by ionization damage, such as the oxide-trapped charges and interface traps, have a significant impact on the performance of a CCD sensor in DUV. The interaction of DUV radiation with CCDs is studied in Chap. 9 to Chap. 12.

7.2 Radiation-Induced Defects in Si, SiO_2, and Si-SiO_2 Interface

In electronic materials, defects attract attention because they introduce a localized electronic level in the forbidden band-gap. Often, the major defects created during radiation, such as vacancies and interstitials, are not stable at room temperature. The defects can interact with each other and with the impurities to form a variety of complex defects (e.g., divacancies and impurity-vacancy pairs) following a room temperature irradiation. The nature of the radiation-induced defects depends on several parameters such as the spatial distribution of the primary defects, the type and concentration of the impurities, and the Fermi level position. In this section, the radiation-induced defects in the Si, SiO_2, and interface layers of the Si-SiO_2 system are reviewed.

7.2.1 Radiation-Induced Defects in Si

Radiation can cause ionization damage and displacement damage in Si, which result from the broken bonds and the atomic displacement in the lattice, respectively. In the case of the broken bonds in the Si lattice, the dangling bond state results in a localized level in the middle of the Si band-gap; atomic displacement usually results in vacancy-related defects in the Si lattice. Radiation-induced defects in Si manifest themselves in the following forms: interstitials, vacancies, divacancies and larger vacancy clusters (e.g., trivacancy, quadrivacancy), vacancy-oxygen pairs, vacancy-doping atom pairs, and vacancies associated with oxygen and carbon [41].

The interstitials in Si are mobile at all temperatures (at least under irradiation), and can interact with Group III dopants to give rise to interstitial impurities. Vacancies are also mobile, but only at temperatures above 80 K [41]. The mobile vacancies interact with each other to form divacancies first, and then form larger vacancy clusters. Vacancies can also interact with various impurities to generate different forms of defects.

The distribution of these radiation-induced defects in Si depends on the type of radiation. For ionizing radiation, the resulting vacancies and interstitials are usually distributed in a highly non-uniform fashion; in this situation, vacancy clusters are the dominant defect species, since their local concentration is very large compared to the concentration of the other impurities. For radiation of energetic particles such as γ-rays or electrons, the resulting vacancies and interstitials are more uniformly distributed [41].

7.2.2 Radiation-Induced Defects in SiO_2

The most frequently identified intrinsic point defects that are induced by radiation in SiO_2 include oxygen vacancy species such as E' centers ($\equiv Si^{\bullet}$), NBOHCs ($\equiv Si - O^{\bullet}$), peroxy radicals ($\equiv Si - O - O^{\bullet}$), and self-trapped holes ($\equiv Si - O^{\bullet} - Si \equiv$) [41, 43, 44, 49].[2] These radiation-induced point defects can participate in different charge transfer processes, as well as contribute to changes in the material absorption. Of the various defects created in SiO_2, radiation-induced oxide-trapped charges and hole traps are the two major threats to the electrical performance and the stability of Si-based devices. The formation mechanisms of these defects will be examined next.

When SiO_2 is exposed to ionizing radiation with a $E_{\rm ph}$ that is greater than the $E_{\rm G}$ of SiO_2, a net positive trapped charge can build-up in the oxide region of the Si-SiO_2 structures [49]. Ionizing radiation creates e-h pairs in the oxide by breaking Si-O bonds. Although some of the photogenerated charge carriers recombine, most drift in the direction of the applied (or built-in) electric field towards the appropriate electrode. Electrons, with their higher mobility, rapidly drift toward a positive electrode, and flow into the external circuit.

---

[2] Refer to Chap. 5 for a description of these defects in SiO_2.

7.2 Radiation-Induced Defects in Si, SiO$_2$, and Si-SiO$_2$ Interface

Also, because thermally grown oxides generally have a low concentration of electron traps, almost all the electrons exit the oxide region. However, since holes have a lower mobility, they drift more slowly toward the interface under the electric field. A fraction of the holes are trapped in the Si-SiO$_2$ interface region (the hole trap distribution usually extends a few nanometers from the Si-SiO$_2$ interface). The rate of the accumulation of trapped holes at the interface depends strongly on the bias conditions during irradiation.[3]

Also, if the free holes drifting in the oxide become trapped by the oxide defects, oxide-trapped charges are created. Oxygen vacancy defects are a major source of the radiation-induced oxide-trapped charges and the hole traps in SiO$_2$; the E' center is the most common under this category. The E' center, $\equiv$Si$^\bullet$, behaves as a dominant deep hole trap in a wide variety of gate oxides, and its presence is often associated with a trapped hole near the Si-SiO$_2$ boundary to initiate oxide charging [49]. The creation of an E' center in SiO$_2$ by radiation was described by (5.6) which was

$$\equiv \text{Si} - \text{Si} \equiv \xrightarrow{Irradiation} \equiv \text{Si}^\bullet + {}^+\text{Si} \equiv + \text{e}^- , \quad (7.3)$$

where the defect precursor before irradiation is an oxygen vacancy, $\equiv$Si$-$Si$\equiv$, located near the Si-SiO$_2$ interface. Also, the E' centers are formed via other radiation-induced reactions, as described in Sect. 5.2. The formation of radiation-induced oxide-trapped charges and hole traps can originate from impurities as well, including OH, H, Na, Cl, and N in the oxide regions near the Si-SiO$_2$ interface [49].

Studies have proven that the density of radiation-induced oxide-trapped charges, N_{ot}, varies superlinearly with oxide thickness [49]. This relationship suggests that N_{ot} can limit the radiation hardness of a device, especially in regions with thick oxide layers (e.g., the field oxide regions used for electrical isolation in MOS circuits). The fact that N_{ot} depends on the oxide thickness, and that the oxide charges affect the flat-band voltage of CCD cells, allude to a dependence of CCD parameters (e.g., the QE and responsivity) on the oxide thickness. This possibility is considered in Chap. 12.

There are several post-radiation recovery techniques to reduce the radiation-induced trapped charges in the SiO$_2$. One approach is to irradiate subband-gap photons (e.g., 4.2 eV of UV light). These photons are incapable of creating e-h pairs in the oxide but is capable of exciting (or photoemitting) electrons from the Si into the SiO$_2$ for recombination, which annihilate the positive trapped charges. The oxide-trapped charges can also be suppressed by thermal annealing.[4]

---

[3]The influence of these trapped charges on MOS devices is examined in Sect. 7.3.1.

[4]Radiation-hardening and annealing techniques are presented in Sect. 7.3.3.

7.2.3 Radiation-Induced Defects in Si-SiO$_2$ Interface

The defects that are induced by irradiation can modify the interface state density of a Si-SiO$_2$ structure. Radiation-induced interface states are classified into three categories depending on their origin. The first category includes the defects that are created in the bulk material (Si or SiO$_2$), but are located close to the Si-SiO$_2$ interface so that they interact with the interface. These bulk defects usually introduce deep defect levels in the energy band-gap. Because they are distributed at random distances from the interface, they contribute to the density of interface states as broad bands in the band-gap than as localized states [41].

The second category entails defects that are directly formed in the interface region as a result of collision and ionization processes. These radiation-induced interface states can be of intrinsic origin such as self-trapped charges, bond rearrangement, or trapping of carriers at a strained interface bond; or the interface states can be extrinsic defects that are associated with impurities. For instance, a charge trapped in an impurity at the interface can modify the atomic configuration of this impurity and of the atoms which surround it [41]. Also, light impurity atoms (such as hydrogen) are displaced by radiation to result in the formation of dangling bonds which contribute to impurity-related interface states.

Finally, the last category constitutes the interface defects that originate from the bulk, but migrate and become trapped in the interface region. These interface-trapped charges saturate at a higher total radiation dose than that of the oxide-trapped charges. Studies have revealed that the generation of interface traps by ionizing radiation are reduced by applying compressive stress at the Si-SiO$_2$ interface [41].

7.3 Effects of Radiation on Basic Semiconductor Devices

The majority of semiconductor devices contains SiO$_2$ or silicon nitride (Si$_3$N$_4$) layers for passivation or isolation purposes. Since these insulating layers are especially prone to charge accumulation from ionizing radiation, the post-irradiation behavior of these devices are often dominated by ionization damage. The failure rate of the irradiated device is strongly dependent on the technological process, the bias during irradiation, and the dose rate. At low dose levels, ionization damage in the passivation and insulating layers is predominant. Typically, ionization damage generates charge defects in the oxide and the interface that alter the electrical characteristics of Si-based devices. However, the ensuing variations of the device's electrical parameters tend to saturate as the absorbed dose is increased. At higher dose levels, displacement damage in the Si substrate becomes predominant. This tends to startle devices that are sensitive to variations in excess carrier lifetime. At extremely high dose rates, transient ionization effects are exhibited, and the devices are liable for the logical set-up and latch-up phenomena in ICs [41].

In this section, the effects of radiation on the operation of MOS structures and electro-optical devices, the annealing of radiation damages, and the radiation hardening of semiconductor devices are discussed. Since the material and processing of CCD sensors resemble those of the MOS technology, similar radiation-induced effects are anticipated in CCDs, and are examined in Chap. 8.

7.3.1 Radiation Effects on MOS Structures

The two primary radiation effects on MOS structures are the charge build-up in the insulating layers, and the increase in Si-SiO$_2$ interface states. Because MOS devices are surface-effect devices, they are very sensitive to changes occurring at the Si-SiO$_2$ interface, in the gate oxide, or even in the upper insulating layers. Typically, the oxide layer acquires a net positive charge after irradiation. But, the charge state of the interface states after irradiation depends on the bias that is applied to the device because interface states can exchange their charges freely with the Si substrate. These radiation-induced changes in the electrical state of the oxide and interface layers affect device parameters. For example, the threshold voltage, the transconductance, and the leakage current of MOSFETs are highly susceptible to radiation-induced modifications [49]. Moreover, radiation can generate a large flux of e-h pairs in the Si, which is responsible for the latch-up effects or soft errors in CMOS (complementary-MOS) circuits.

The radiation effects on MOS structures involve four stages that occur in different time scales after irradiation, as illustrated in Fig. 7.2. The first stage is the initial hole yield, where the ionizing radiation interacts with the SiO$_2$ to produce e-h pairs. Depending on the kind of radiation and the magnitude of the applied electric field, these carriers can recombine within the oxide or they are separated and transport through the oxide at distinct speeds. The mobile electrons are swept out of the oxide by the applied field. In contrast, the holes have a very low effective mobility and are transported via a complicated stochastic trap-hopping process toward the Si-SiO$_2$ interface. This hole transport process constitutes the second step in Fig. 7.2 [43].

The third stage is identified as the hole trapping process. Some of the holes are easily trapped in the oxide, leading to the build-up of radiation-induced oxide charges. Other holes drift to the Si-SiO$_2$ interface, where they capture electrons and create interface traps. The number of free or untrapped holes dictates the response of the semiconductor devices. For example, when the holes become trapped in the oxide, the oxide layer acquires a net positive charge and causes a shift in the threshold voltage, $\Delta V_\mathrm{T}(Q_\mathrm{ot})$, where Q_ot is the positive trapped charge in the oxide. This $\Delta V_\mathrm{T}(Q_\mathrm{ot})$ usually saturates with the radiation dose and is due to three circumstances. First, since the number of available neutral traps decreases as the dose is increased, the

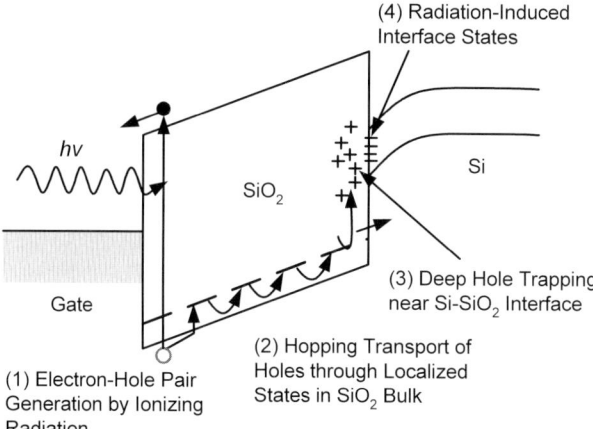

Fig. 7.2. The processes of basic radiation effects in a MOS structure (adapted from Nishikawa [43])

probability of trapping is reduced. Secondly, an increase in the electric field in the charge accumulation region weakens the electric field in the remaining bulk of the oxide; consequently, there is an enhanced recombination process in the field-free oxide region which lowers the rate of charge build-up. Lastly, the saturation is a consequence of the electron trapping in the now positively charged traps [41].

Since the hole mobility varies greatly with temperature, different radiation-induced effects evolve in the MOS structure depending on the temperature. At low temperatures, the holes are relatively immobile and the density of the positive charges (both immobile and trapped) is uniform throughout the oxide. However, at room temperature, the radiation-generated holes are mobile in the presence of an electric field in the oxide, and only those holes which are capable of reaching the vicinity of the Si-SiO$_2$ interface have a probability of being trapped. Since the distribution of trapped holes is no longer uniform, it becomes more difficult to predict the radiation-induced shifts in the MOS characteristics as the temperature is increased.

The fourth stage of the radiation effects in MOS structures is the build-up of radiation-induced interface states. A number of models have been proposed to account for the build-up of interface states. One such model involves the transport of radiation-generated free hydrogen ions (H$^+$) through the SiO$_2$, where the liberated hydrogen ions undergo a dispersive hopping transport that controls the rate of the interface state generation. An alternate model has demonstrated that the arrival of the radiation-induced holes at the interface and the diffusion of neutral hydrogen are responsible for the interface state build-up [43]. The increase in the interface state density, ΔD_{it}, after radiation can inflict $\Delta V_T(\Delta D_{it})$ in the MOSFETs. Depending on the type of interface states (acceptor or donor types), the interface charge can

7.3 Effects of Radiation on Basic Semiconductor Devices

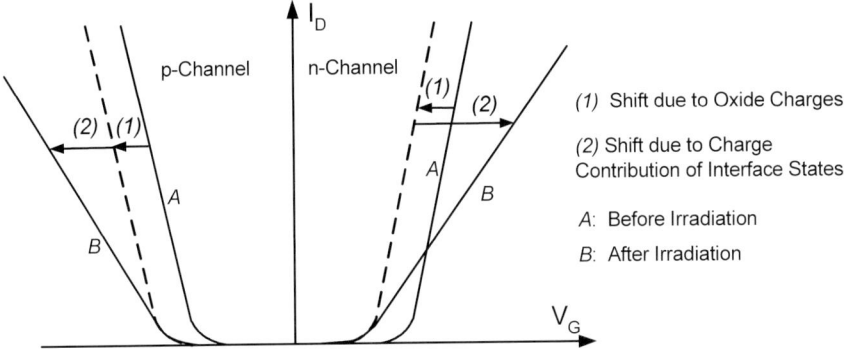

Fig. 7.3. The effects of radiation-induced oxide charges and interfacial charges on the I_D-V_G characteristics of p-channel and n-channel MOSFETs: (1) the negative shift due to oxide charges, and (2) the shift due to charge contribution of interface states. The increase of the interface state density results in a decrease of the forward transconductance (adapted from Barbottin et al. [41])

augment or compensate for the total threshold voltage shift, as given by $\Delta V_{T,\text{total}} = \Delta V_T(Q_{\text{ot}}) + \Delta V_T(\Delta D_{\text{it}})$ [41]. Moreover, the creation of interface states signifies that more interface trapping will occur, causing a degradation of the carrier mobility in the channel of a MOS transistor. This leads to a reduction in the channel conductance and transconductance, and thus, a decrease in the transistor gain [41].

For low radiation doses, the ΔV_T in MOSFETs is initially due to the oxide-trapped charges. For intermediate doses, the generation of a compensating interface charge becomes possible. The dependence of the oxide charge build-up and the interface state generation on the radiation dose is a potential reason for the different radiation-induced dark current behavior observed when the CCD sensors are exposed to varying DUV laser doses.[5] On some occasions, the radiation-induced ΔV_T can recover after the irradiation period; however, the recovery process, which is determined by the hole transport and trapped-holed annealing mechanisms, can extend from milliseconds to years.

In addition, the radiation-induced oxide-trapped charges and interface states cause distortions in the C-V characteristics of a MOS device. The oxide-trapped charges shift the C-V curve in the negative direction, and the interface traps stretch or expand the C-V curve. Consequently, a greater change in the applied bias voltage is required to create the same change in the capacitances as prior to the irradiation [49].

The type of channel doping also affects the radiation tolerance of a MOSFET. The effects of radiation on the current-voltage, I_D-V_G, characteristics of the p- and n-channel MOSFETs are illustrated in Fig. 7.3. The two main effects are a shift in V_T, and a lowering of the transconductance ($= \partial I_D/\partial V_G$).

---

[5] This mechanism is examined in Chap. 12.

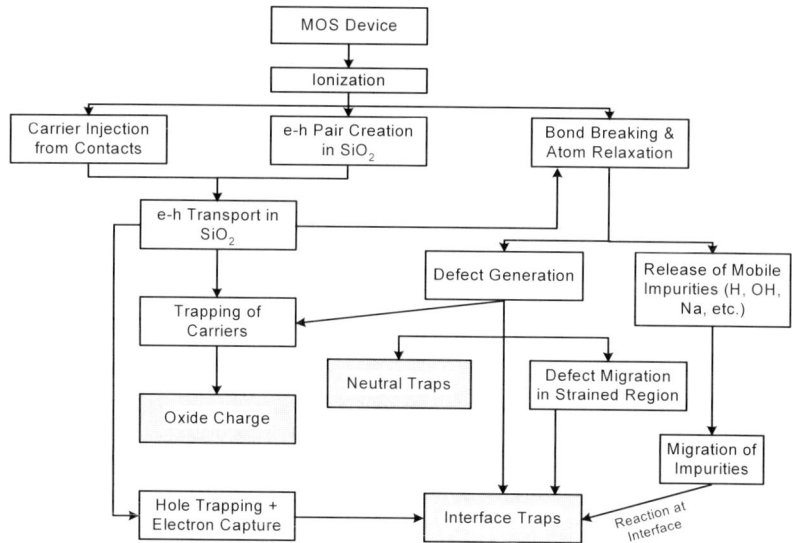

Fig. 7.4. A summary of the possible processes by which ionizing radiation leads to the creation of oxide charge, neutral traps, and interface traps in a MOS device (adapted from Ma and Dressendorfer [49])

Under identical biasing conditions, the radiation-generated interface states induce a more negative ΔV_T for the p-channel transistor than for the n-channel transistor [41]. In general, the n-channel devices are more sensitive to radiation effects than the p-channel devices. If the radiation-induced oxide charges cause the I_D-V_G curve to shift in a negative V_G direction, then a significant leakage current can flow in an n-channel device even if $V_G = 0$. This implies that the n-channel transistor cannot be turned off efficiently after irradiation. This leakage current in the n-channel transistor causes excessive power dissipation in the CMOS circuits. However, the same does not apply to the p-channel transistor.

Figure 7.4 is a summary of the effects of ionizing radiation on MOS devices [49]. The flowchart clearly demonstrates that the primary consequence of ionization damage is the creation of oxide charges, neutral traps, and interface traps. Similar radiation processes occur in CCDs, and will be considered in Chap. 8. For a particular type of ionizing radiation, the radiation effects that are presented in this section are applicable, but more specific damages can emerge as well. For example, DUV laser radiation not only causes an increase in the oxide charges and interface states, but also other damages, such as the DUV-induced color center formation and the induced absorption in SiO_2, are often observed.[6]

---

[6]DUV-induced effects are followed up in Chap. 10.

7.3.2 Radiation Effects on Electro-Optical Devices

Si-based electro-optical devices exhibit a variety of radiation sensitivities, depending on the design, structure and operation of the device. For example, photo-transistors are very sensitive to ionizing radiation. The increase of radiation-induced dark current, due to the creation of interface states, can become significant. Also, the transistor gain and responsivity can be severely degraded due to the radiation effects on the photo-transistors. Guard rings are used to lessen the radiation damage of Si photo-transistors and photocells [41].

CCD image sensor is another example of a Si-based electro-optical device. CCDs are also sensitive to ionizing radiation, especially if the device is operating at low temperatures. Buried channel CCDs have a superior radiation tolerance compared to that of surface channel CCDs. This is attributed to the extra implants in the buried channel devices, which separate the active channel area of the CCD from the Si-SiO$_2$ interface, rendering the active region less sensitive to the creation of radiation-induced interface states. A more thorough examination of the radiation effects on CCDs will be presented in Chap. 8.

7.3.3 Annealing of Radiation-Induced Defects

Annealing is performed in an attempt to restore the device performance and/or to relieve the radiation-induced damages after a device has been exposed to radiation. In the case of displacement damage, annealing helps to restore the lattice structure of the crystal to its pre-irradiation condition.

The annealing process alters the microstructures of a material and annihilates the defects, and can be accomplished by the diffusion of defects to the surface, the recombination of interstitials and vacancies, or the re-orientation of defects [41]. Various types of annealing processes are available, including thermal annealing, current injection annealing, and laser annealing processes, which are thermally, electrically, or optically activated, respectively.

7.3.4 Radiation Hardening

The radiation hardness of an electronic device is defined as the level of radiation exposure that can be tolerated before the device performance degrades to an unacceptable level. By radiation hardening a device, it becomes less sensitive or vulnerable to radiation damage. Hardening procedures vary, depending on such things as the type of circuit, its operating range, the type of radiation environment, and the anticipated failure mechanism.

In the case of displacement damage, the main failure mechanism is a reduction in the lifetime, τ. This is usually unavoidable, and radiation hardening techniques are not very applicable to address displacement damage issues. The resulting loss in function due to displacement damage has to be accounted for the during the circuit design [41].

In the case of ionization damage, the inherent sensitivity of MOS devices can be reduced by technological means. For instance, a simple way to enhance the radiation hardness of CMOS circuits is to replace n-channel devices with p-channel devices, since the n-channel devices are more vulnerable to radiation damage. To reduce the sensitivity of MOS devices to transient ionization bursts, the active area of the device must be kept small and any leakage paths must be suppressed by insulating barriers.

The key parameters to consider when designing a radiation hardening procedure for a MOS device are the hole mobility (which is strongly related to the processing steps), and the hole capture rate in traps located near the Si-SiO$_2$ interface [41]. These two parameters directly influence the radiation-induced oxide-trapped charges, interface states, and hole trapping. In addition, the oxide thickness, t_{ox}, and the processing conditions determine the radiation hardness of a device. In terms of the oxide thickness, studies have indicated that for a dry radiation-hardening process, the radiation-induced ΔV_T (at room temperature) has a t_{ox}^3 dependence regardless of the applied bias during irradiation. Since geometrical considerations only account for a t_{ox}^2 dependence, the additional factor is related to the process (e.g., the hole transport and hole capture processes) [41]. Therefore, radiation hardness is improved by reducing t_{ox}. However, a reduction in t_{ox} can conflict with other aspects of the device reliability; for instance, if the gate oxide is too thin, it becomes more susceptible to dielectric breakdown. Thus, t_{ox} must be carefully chosen to ensure the MOS device maintains good electrical reliability and radiation hardness.

The radiation hardness of a device is also influenced by the processing conditions (e.g., the oxidation process and temperature). For example, HCl steam oxidation improves the radiation hardness of a device; also, an oxidation temperature of 1000°C is optimum for radiation hardening [41]. All post-oxidation high temperature process steps need to be performed at temperatures lower than the oxidation temperature to maintain the material properties and the device integrity. Thus, the influence of other processing steps (such as post-oxidation annealing, gate-deposition, and metallization) need to be thoroughly assessed to produce a device with superior radiation hardness.

Since CCD sensors are intended to operate in a radiation environment, radiation hardening of CCDs is very important. The effects of radiation on CCDs and the parameters that affect the radiation hardness of CCD sensors are examined in the next chapter.

8 Effects of Radiation on CCDs

Incident photons can interact with CCD sensors in one of the three mechanisms: the photoelectric effect, Compton effect, and pair production.[1] In all three cases, the interaction produces free electrons that can contribute to the CCD response. Low-energy photons (approximately <0.1 MeV) interact through the photoelectric effect. Here, the incident photon excites and emits a valence electron, ionizing the target atom. If the emitted electron has sufficient energy, it excites the neighboring atoms to generate additional e-h pairs.

Typically, the energy required to create a single e-h pair in Si is 1.1 eV. More energetic photons can generate multiple e-h pairs, where one e-h pair is generated for every 3.70 eV absorbed in Si [14,49]. For SiO_2, the minimum energy that is required to create a single e-h pair is approximately 9 eV ($\lambda =$ 138 nm). This implies that photons with an energy greater than or equal to 9 eV are absorbed by the oxide layer and direct damage to the oxide layer begins to take place. The average energy required to create multiple e-h pairs in SiO_2 is 18 eV [14]. Thus, the photogeneration of e-h pairs in SiO_2 is possible with DUV, VUV, X-ray, and γ-ray radiations; as such, the CCD is susceptible to ionization damage at these wavelengths. Since the interest in this book lies in CCD sensors for DUV imaging, the ionization damage in SiO_2 is a relevant concern, and must be dealt with. This chapter begins with an overview of the general radiation damages that are encountered by CCDs. This is followed by a more in-depth discussion on the mechanisms of the ionization damage in CCD sensors.

8.1 Overview of the Radiation Damages in CCDs

CCD image sensors are prone to the two major radiation damage mechanisms that were described in Sect. 7.1: ionization damage and displacement (or bulk) damage. Figure 8.1 signifies the regions of a CCD sensor that are susceptible to radiation damage and the associated damage mechanisms. Typically, radiation damage in CCDs leads to the creation of new Si-SiO_2 interface states, the generation of charge species in the oxide, and atom

---

[1] Refer to Sect. 7.1.3 for a description of these mechanisms.

110   8 Effects of Radiation on CCDs

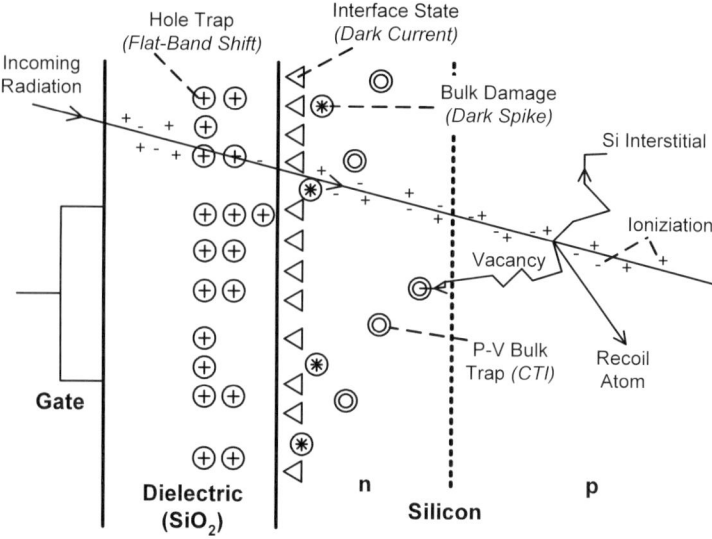

Fig. 8.1. The ionization and bulk damage effects occurring at the SiO_2 dielectric and epitaxial Si layer of a CCD sensor. The gate structure may or may not be present, depending on the type or design of the CCD sensor. For example, the gate electrode is absent in the active imaging region of a photodiode-based CCD (adapted from Janesick [14])

displacement. These physical changes subsequently trigger an increase in the dark current, flatband voltage shifts in the CCD amplifiers, long term degradation of the CTE, and a degraded image quality. Since the photosensitive region (or imaging region) of a DUV-sensitive CCD cannot have any overlying polysilicon layer (polysilicon has a high DUV absorption), this investigation is concentrated on the radiation damage effects that are associated with the SiO_2 layer and Si-SiO_2 interface of a CCD sensor.

Similar to that of the MOS devices, the oxide layer of the CCD is the most sensitive to ionization damage. When ionizing radiation passes through the oxide, the energy deposited creates an e-h pair which causes permanent damage to the oxide and interface layers. Two types of damage effects can emerge. First, the holes generated in the oxide layer become trapped; this leads to a charge accumulation at the Si-SiO_2 interface that alters various CCD functions. For example, a charge accumulation can change the capacitance of the photodiode region, which subsequently alters the channel potential and the full-well capacity of a CCD pixel. Another common consequence is the flatband voltage shift, which causes shifts in the bias potentials of the MOS capacitors in the CCD registers and in the output amplifier. If the shift causes the bias potential to fall outside the operating window, the CCD sensor fails to deliver the intended performance. The second effect of ionization damage is the breaking of the weak bonds at the Si-SiO$_2$ interface and the creation interface states. The newly formed interface states generate an additional flatband shift, a surface dark current, and noise [14].

The ionization damage of a CCD is dependent on the type, energy, and rate of the incoming radiation. Energetic photons and electrons are the major ionizing radiation sources. Heavier particles such as protons and heavy ions can also induce ionization damage; however, displacement damage is the principal concern for these types of radiation [14].

Displacement damage in CCDs concerns the epitaxial layer where the charge is generated, collected, transferred, and measured. When energetic electrons, protons, heavy ions, and neutrons pass through the CCD, there is the possibility of collisions with the Si target nuclei. These events displace atoms from their lattice positions and create vacancy-interstitial pairs. The permanent defects that are produced by this process can cause a degradation of the CTE, an increased dark current generation, dark current spikes, and fluctuations in the output amplifier characteristics.

DUV excimer laser irradiation is more likely to introduce ionization damage to CCD sensors. DUV-related damage effects in CCDs will be studied in subsequent chapters. For the remainder of this chapter, the processes and effects of ionization damage in CCDs are considered in greater detail. The processes discussed here are also relevant for the study of DUV-induced damage in CCD sensors.

8.2 Ionization Damage in CCDs

Similar to the ionization processes in MOS devices discussed in Sect. 7.3.1, the study of ionization damage in CCDs examines several processes: e-h generation, e-h recombination, fractional yield, hole transport, hole trapping, annealing, and interface state creation. These mechanisms work together in a complex manner to induce the damages in the CCD's insulating and passivation (SiO$_2$) layers. The discussion in this chapter is focused on the ionization damage effects related to the e-h generation in the SiO$_2$ layer of the CCDs. However, ionizing radiation can induce other defects (e.g., color centers, charged centers, and ionized impurities) that result in additional carrier trapping and charging in the oxide layer; these defects also contribute to the radiation damage in CCD sensors.[2] The tolerance of a CCD sensor to ionization damage varies widely among manufacturers, and is contingent on the processing details and the CCD technology employed; these dependences will also be explored in this chapter.

---

[2]More details on these radiation-induced defects will be examined in Chap. 10.

8.2.1 e-h Generation

Typically, visible and near-IR photons do not produce ionization damage, since they are not absorbed by, and have no interaction with, the SiO_2. SiO_2 begins to absorb photons with a wavelength approaching the DUV region. The first noticeable permanent ionization effect occurs at a wavelength of approximately 290 nm (E_{ph} = 4.3 eV), and is related to the photoemission of carriers [14]. When a CCD is exposed to DUV radiation, a portion of the incident photons pass through the oxide layer and interact with the Si channel. E-h pairs are generated in the Si channel, where the carriers participate in the basic operations of a CCD and contribute to the response of a CCD sensor. However, the incident DUV photons can also interact with the Si channel by photoemitting electrons in the Si into the SiO_2 layer, where the electrons become trapped and induce negative charging of the oxide (i.e., a positive flatband shift). This can lead to shifts in the channel potential and the capacitance of a CCD pixel, which then alter the charge capacity and other electrical characteristics of the pixel.[3]

UV-phosphor coated CCD sensors do not show this photoemission problem because the phosphor coating absorbs all the incoming UV photons and are re-emitted as visible photons toward the CCD active layer. Backside-illuminated CCD sensors are not as vulnerable to ionization damage, because the epitaxial layer acts as a shield to prevent photons that are incident from the backside from reaching the frontside surface; as a result, UV-induced oxide charging is inhibited at the frontside of the sensor. Nevertheless, ionization damage remains a valid concern for backside-illuminated devices, since ionization can occur in the thin native oxide layer that grows naturally on the back surface of the Si substrate. As a result, the DUV photoemission of electrons near the back surface is possible and can cause oxide charging.

For wavelengths shorter than 180 nm, photons can interact directly with the oxide layer and generate e-h pairs in the oxide. True ionization damage is initiated when this happens [14]. The photogenerated carriers in the SiO_2 layer are responsible for the flatband shift and the dark current generation in the CCD sensors from the ionization damage.

8.2.2 e-h Recombination and Fractional Yield

Once e-h pairs are generated in the oxide, they have an opportunity to recombine within a very short time frame, before they are separated by a field-induced drift or by diffusion. If a voltage bias is applied to the CCD, the e-h pairs separate and move in opposite directions due to the electric field in the oxide. In the absence of an applied bias, the carriers migrate mainly through diffusion. Electrons and holes move through the oxide at much different rates.

---

[3] Details on the DUV photoemission of carriers in Si-SiO_2 structures are provided in Chap. 9.

The electron mobility in the SiO$_2$ is 20 to 40 cm^2/(V·s), whereas the hole mobility is considerably less, 10^{-4} to 10^{-7} cm^2/(V·s). A slight bias causes the electrons to move near their saturation velocity, about 10^7 cm/s [14]. Therefore, if e-h recombination occurs, this process must take place in a very short time period (on the order of a few picoseconds).

After the initial recombination process, a majority of the electrons in the oxide migrates toward the electrode and leaves the oxide; only holes remain in the oxide layer. These holes are the dominant species contributing to the ionization damage in CCDs, since they are involved in the generation of oxide-trapped charges and interface traps. The fraction of the holes that remain in the oxide is described by the fractional yield, h_{FY}, as follows:

$$h_{\text{FY}} = \frac{N_{\text{h}}}{N_{\text{e-h}}}, \tag{8.1}$$

where N_{h} is the number of remaining holes, and $N_{\text{e-h}}$ is the initial number of e-h pairs that are generated from the incident radiation [14]. The fractional yield is strongly dependent on the energy and type of incident particle, and increases rapidly as a bias is applied to the CCD because the resulting electric field promotes the e-h separation.

8.2.3 Hole Transport

When the holes separate from the electrons in the oxide, the holes either move to the Si-SiO$_2$ interface or to the gate electrode (if present), depending on the direction of electric field. The hole transit time depends on the electric field strength, operating temperature, and thickness of the oxide layer. For example, at room temperature, hole migration is on the order of 10^{-3} s for a field strength of 1 MeV/cm. However, if the CCD is very cold and unbiased, a time duration on the order of 10^3 s is necessary for a complete hole transport [14]. During the transport process, the holes can be trapped by the defects in the oxide, or by the defects at the Si-SiO$_2$ interface; thus, hole trapping must be considered.

8.2.4 Hole Trapping

Immediately after an irradiation, the holes that are generated in the oxide cause a net negative flatband voltage, V_{FB}, given by

$$V_{\text{FB}} = -\frac{1}{\varepsilon_{\text{ox}}} \int_0^{t_{\text{ox}}} x \rho_{\text{ox}}(x) dx, \tag{8.2}$$

where t_{ox} is the thickness of the oxide, $\rho_{\text{ox}}(x)$ is the distribution of holes in the oxide, and x is the distance of an arbitrary location in the oxide layer from the gate electrode. Equation (8.2) indicates that the largest flatband voltage shift occurs when the holes are trapped at the Si-SiO$_2$ interface (i.e., when x is large and approaches t_{ox}); in fact, this is predominantly the case in real life [14]. A flatband shift is highly undesirable because it can cause fluctuations in the device parameters with the radiation dose and induce device instabilities (e.g., the V_T shifts, and the changes in charge capacity of a pixel).

It is noteworthy that the radiation-induced positive flatband voltage shift can also occur, as a result of the generation of the negatively-charged defects in the oxide layer. The net flatband voltage shift of a CCD pixel is governed by the relative influence of the negative oxide charges and the hole trapping in the oxide layer.

8.2.5 Annealing (Detrapping of Holes)

The trapped holes in the dielectric are not permanent features, since they can slowly detrap (i.e., anneal). Two mechanisms that anneal a hole trap are tunnel annealing and thermal annealing [14]. Tunnel annealing occurs when an electron from the Si layer tunnel into the SiO_2 layer and recombine with a trapped hole. Tunnel annealing tends to reduce the amount of positive charge trapped in the oxide. Thermal annealing occurs when a hole escapes from a trap, owing to thermal excitation. The speed at which these processes work depends on the physical distance of the hole trap from the Si-SiO_2 interface, and the emission time constant of the trap (related to E_T). Therefore, after irradiation, the flatband voltage can take days to stabilize and require even longer periods at cold operating temperatures [14]. The trapping and detrapping of holes in the oxide can cause fluctuations in the device characteristics, resulting in poor device stability.

8.2.6 Interface State Creation

Another consequence of ionization damage in a CCD is the creation of interface states, which was discussed in Sect. 7.3.1 for the case of MOS structures. Interface states introduce energy levels inside the forbidden band-gap. Good interfaces are characterized by an interface state density between 10^9 and 10^{11} traps/cm^2-eV at the mid-band [14]. These mid-band states play an important role for dark current generation. The actual interface state density is dependent on the fabrication details and varies significantly among manufacturers. Additional interface states can be generated by radiation, which subsequently lead to an increase in the dark current, a shift in the flatband voltage, and a degradation of CTE and QE.

A possible form of the radiation-induced interface state in a CCD is the P_b center dangling bond. A significant portion of dangling bonds can be passivated by the incorporation of hydrogen (H) into the CCD process. This process is described by

$$\equiv \text{Si}^{\bullet} + \text{H} \longrightarrow \equiv \text{Si} - \text{H}, \tag{8.3}$$

where $\equiv$ Si$^\bullet$ represents a dangling bond (i.e., an unpaired electron) at the interface. Usually, this hydrogen annealing technique is employed throughout the CCD fabrication process. It is especially important when the fabrication involves high-energy plasma etching step(s), because the high-energy radiation from the plasma can introduce an appreciable number of interface states. However, a drawback of using hydrogen passivation is that the CCD can become more vulnerable to ionization damage [14]. Radiation can break the weak passivating Si-H bonds and form dangling bonds, which then cause a flatband voltage shift and generate surface dark current. Thus, the hydrogen passivation of interface states is not suitable for the radiation-hard CCD sensors.

Interestingly, for some CCDs, dark current and flatband voltage can continue to increase for several years after the sensor is irradiated. This long-term reaction is called *reverse annealing*, because the flatband shift and dark current increase, rather than decrease, after the CCD is irradiated [14]. Reverse annealing is not associated with the simple hole transport process, because the hole movement is too rapid, in the order of a few seconds at room temperature, to be accountable for the device changes that occur on a monthly or yearly time scale. Models that are based on interface state generation appear to be more appropriate for explaining the reverse annealing effect.

One model of reverse annealing considers a multi-step process for the creation of new interface traps. The first step is the transport of the radiation-induced holes through the oxide toward the interface. Next, the holes interact and break the weak bonds between a H atom and a trivalent Si atom in the oxide. An H$^+$ ion is released from this reaction, and drifts toward the Si-SiO$_2$ interface. On reaching the interface, the H$^+$ ion picks up an electron to form a neutral H atom. This H atom then reacts with a hydrogen-passivated bond at the Si-SiO$_2$ interface, producing H$_2$ and a dangling Si bond. The resulting dangling bond contributes to the reverse annealing processes. In equation form, the last two reactions are described by

$$3H - Si - H + H^+ + e^- \longrightarrow 3H - Si - H + H^0 \longrightarrow 3H - Si^\bullet + H_2 \,, \quad (8.4)$$

where 3H−Si−H is a hydrogen passivated trap, and 3H−Si$^\bullet$ is the new dangling bond (or interface trap). The diffusion process of an H$^+$ ion in the oxide is exponentially dependent on the operating temperature. It can take several hours for this reaction to stabilize at room temperature. However, the reaction time is reduced to a few minutes at elevated temperatures (>50°C) [14]. Reverse annealing is dependent on the bias condition, since holes and ions are charged species. For example, when a CCD gate is negatively biased relative to the substrate, the resulting electric field attracts the H$^+$ ions to the gate instead of the interface; in theory, this stops the reverse annealing. Contrarily, a positive bias on the CCD gate repels the ions to the interface, accelerating the reverse annealing reaction. Therefore, reverse annealing can be controlled by the applied bias and temperature.

8.2.7 Dependence of Ionization Damage on Insulator Properties

Analogous to MOS structures, the oxide thickness and oxidation conditions affect the radiation hardness of CCD sensors.[4] A radiation-tolerant hard oxide and a radiation soft oxide are distinguished by the thickness of the oxide, the manner in which the oxide is grown (e.g., wet oxidation or dry oxidation), and the Si-SiO$_2$ interfacial stress that is induced by the overlaying gate or passivation layers. Of these three factors, the oxide thickness is the most important because the number of e-h pairs generated is proportional to the volume of the oxide layer [14]. In theory, if the oxide layer is thin enough, there is no particle interaction and ionizing damage in the oxide. Devices with thin oxides have very little flatband voltage shift, because the oxide layer can be neutralized by tunnel annealing.[5]

The choice of insulator material also has an important impact on the ionization damage of CCDs, and the most common insulator materials are SiO$_2$ and silicon nitride (Si$_3$N$_4$). If Si$_3$N$_4$ is used, trapping in the insulator appears to be negligible because most of the e-h pairs in the Si$_3$N$_4$ layer either recombine or are trapped very close to the generation site. Thus, the instabilities, that are related to hole trapping in the insulator layer, are of less concern in the Si$_3$N$_4$ layers. However, Si$_3$N$_4$ tends to introduce a larger interface state density at the semiconductor-insulator interface.

Most CCD and MOS devices are composed of regions of thick field oxide (typically 10 000 Å) over the channel stop regions for isolation, and thin gate oxide (typically 500 Å) over the signal channel [14]. These two regions converge at an interface called the bird's beak. Since the field oxide is approximately 20 times thicker than the gate oxide, more radiation damage is expected to occur here. As the bird's beak is approached from the signal channel, the oxide thickness increases which results in more radiation damage and a greater flatband shift. Also, the channel potential increases as the bird's beak is approached due to a possible positive charge build-up in the field oxide. As a result, it takes a larger negative gate voltage to maintain the channel inversion near the bird's beak region, than the signal channel in CCDs does. In addition, a flatband shift in the bird's beak causes a degradation in the full well capacity of the pixel. Thus, careful design strategies are needed to compensate for the varying degrees of radiation damages in the different regions of the CCD sensors. One possible CCD design technique that alleviates the radiation problem at the bird's beak is the super notch structure. The super notch isolates the bird's beak by leaving a small neutral p-type region between the signal channel and the channel stop to support an inverted state in the signal channel region [14].

---

[4] The effect of oxide thickness and oxidation condition on the radiation hardness of a MOS device was discussed in Sect. 7.3.4.

[5] The significance of the oxide thickness on the DUV response and on DUV-induced damage of CCDs are considered in Chap. 12.

8.2.8 UV Flood

It is possible to reverse some of the radiation-induced negative flatband shifts by using the UV flood technique. This technique exploits the concept of photoemission of electrons into the oxide layer to help neutralize the positive charge build-up in the oxide layer [14]. A UV flood can significantly suppress the increase in dark current due to ionization damage, and cure the radiation-induced degradation in CTE. The technique is applied routinely for backside-illuminated CCDs, where the photoemission of electrons into the native oxide layer on the backside neutralizes some of the positive oxide charging effects.

Part IV

Interaction of UV Radiation with the Si-SiO$_2$ System

9 UV-Induced Effects in Si

There are various radiation-induced phenomena that contribute to the degraded performance and instability of CCD sensors from UV irradiation. As illustrated in Fig. 9.1, UV photons have sufficient energy to access a number of energetic processes in a Si-SiO$_2$ system. For example, the photoemission of electrons from Si to SiO$_2$, and the dissociation of O$_2$ molecules are possible with UV excitation. In addition to these processes, UV radiation triggers a number of complex mechanisms and defect formation reactions in the Si-SiO$_2$ system, including the UV-induced absorption in SiO$_2$ due to color center formation, UV-induced oxide charging, structural rearrangement in the SiO$_2$ layer, and interfacial modification. These phenomena are explored in Chap. 10 and Chap. 11.

To analyze the effects of DUV excimer laser irradiation on the behavior of a CCD sensor, it is important to first recognize the interaction of DUV laser photons with the CCD material system. Therefore, Part IV of this book is dedicated to examining the UV laser-induced effects in the key materials that are present in the photosensitive region of a CCD sensor, which includes Si, SiO$_2$, and the Si-SiO$_2$ interface.[1]

9.1 Photoemission in Si

Figure 9.1 shows that a number of energetic processes in a Si-SiO$_2$ structure is activated by UV radiation. In addition to the photogeneration of e-h pairs in Si, UV photons can cause the photoemission of carriers in the Si into the SiO$_2$ layer. The investigation by William [51] and Goodman [52] have revealed that the transition of electrons from the Si conduction band (CB) to the SiO$_2$ CB is possible by an excitation of 3.15 eV photons; the transition of electrons from the Si valence band (VB) to the SiO$_2$ CB occurs with a photoemission threshold of 4.25 eV at room temperature [51]. These photoemission thresholds correspond to the UV wavelengths of 388.24 nm and 288.9 nm, respectively. The photoemission of the holes from the Si VB to the SiO$_2$ VB can also take place with a threshold energy of 4.6 eV [52].

---

[1] The discussion of UV-induced effects encompasses DUV-induced effects, since DUV is a subset of UV.

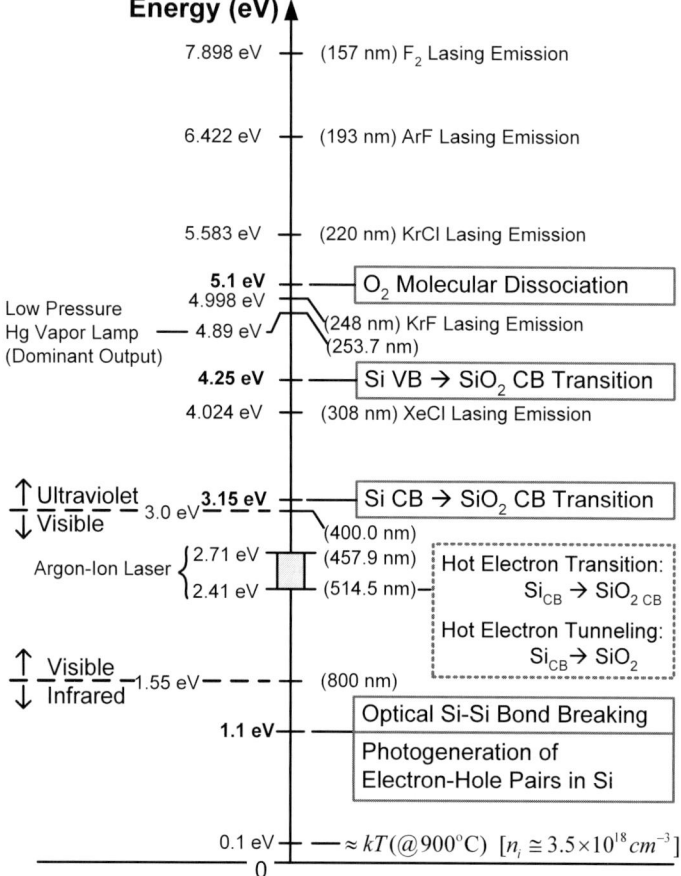

Fig. 9.1. The energetic processes, in a Si-SiO$_2$ structure, that are accessible to IR, visible and UV radiations. The sources of radiation at the different UV wavelengths are also indicated (adapted from Young and Tiller [50])

This energy can be supplied by DUV photons with a wavelength shorter than 270 nm. Using these photoemission threshold data and assuming the optical absorption edge of SiO$_2$ is approximately 8.9 eV, an energy band diagram of the Si-SiO$_2$ structure can be constructed and is depicted in Fig. 9.2.

When the photon energy exceeds the photoemission thresholds, the photoinjection of carriers in the Si into the SiO$_2$ increases substantially [51]. The UV-induced photoemission of carriers inevitably affects the performance of the CCD sensors.[2] If electrons are photoemitted from the Si into the SiO$_2$

---

[2] Refer to Sect. 8.2.1 for a discussion of the impact of the photoemission of carriers on the CCD performance.

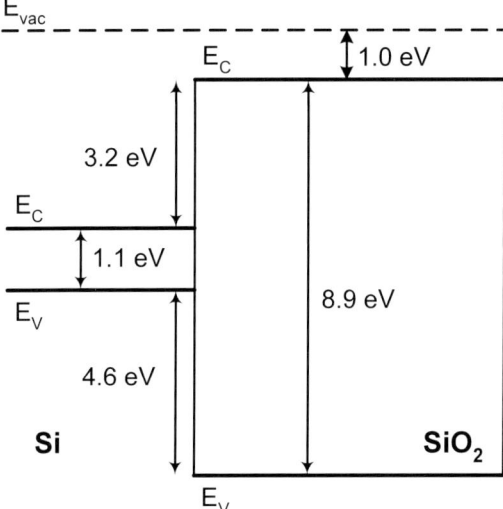

Fig. 9.2. An energy band diagram of a Si-SiO$_2$ structure. The possible bending of energy bands in silicon at the interface is not shown here (adapted from Williams [51])

layer, the loss of the signal electrons in the Si layer leads to a reduction in the CCD photoresponse. Moreover, the photoemitted carriers can be trapped in SiO$_2$ to cause oxide charging and other oxide defects, which subsequently alter the electrical characteristics of the device. As a result, CCD parameters such as dark current, CTE, and QE are susceptible to modifications. The UV-induced effects in SiO$_2$ are discussed in the next chapter.

10 UV Laser Induced Effects in SiO$_2$

It is evident, from the discussion in preceding chapters, that the oxide layer is most prone to radiation damage in CCD sensors. Thus, the majority of the CCD degradation mechanisms at DUV-VUV wavelengths are likely to stem from the ionization damages and the UV-induced effects in the SiO$_2$ layer. The ionization damages in SiO$_2$ were studied in the preceding chapters. In this chapter, the UV laser induced effects in SiO$_2$ are examined. This information lays the foundation for Chap. 12, and establishes a better understanding and justification for the DUV behavior of CCDs.

In addition to being technologically important in the microelectronics sector, SiO$_2$ has unique optical features that are favorable for use as DUV optics (e.g., glasses and lenses) in DUV excimer laser photolithography systems [53]. These features include a high transparency in the DUV-VUV region, an excellent optical homogeneity, and a high durability when the SiO$_2$ is exposed to laser light. Nevertheless, SiO$_2$ is vulnerable to optical modifications as a consequence of prolonged DUV exposure. The DUV photons have sufficient energy to cause permanent property changes in the SiO$_2$, including a degradation in the index homogeneity and optical transmission.

Theoretically, a-SiO$_2$ should have an excellent transparency at DUV wavelengths, since the fundamental absorption edge (given by the band-gap energy) of a-SiO$_2$ is larger than the DUV photon energies [53].[1] However, the transparency of SiO$_2$ is often limited by the pre-existing or radiation-induced intrinsic point defects. The pre-existing defects can be created during the preparation and fabrication process, and most of these defects have characteristic absorption bands in the UV spectrum. The defects and color centers in the SiO$_2$ can also be activated by UV radiation, and often exhibit UV absorption bands to further degrade the UV transparency of SiO$_2$. For instance, defects such as E' centers, NBOHCs, ODCs, and strained Si-O-Si bonds have optical absorption bands in the UV region, and they interact with the UV photons to trigger an assortment of photo-reactions or photo-processes. An example is the photolysis of strained Si-O-Si bonds into the E' center and

---

[1] For simplicity, the term DUV is used to refer to the family of photolithographic wavelengths that includes 248 nm, 193 nm and 157 nm. VUV covers a wider spectrum, and refers to UV wavelengths which require operation in a vacuum. VUV encompasses 157 nm, and wavelengths near and shortward of 157 nm.

NBOHC, which is recognized as an important process in the photodegradation of SiO_2. Also, the distribution and concentration of strained Si-O-Si bonds are known to have a dominant influence on the DUV-VUV absorption of SiO_2 [53]. Evidence of UV-induced processes and damages in SiO_2 is found in the scientific literature, where there has been ongoing research on the damage behavior of bulk SiO_2 glasses for DUV laser applications. The results published in the literature provide a basis for identifying the UV-induced damages in the SiO_2 layer of a CCD sensor.

A distinct feature of the DUV laser induced damage in SiO_2 is the energy absorption process. The DUV excimer laser sources are pulsed; thus although the average power is modest, the peak power of the laser pulses is very high. One consequence of the high peak power is the possibility of two-photon absorption (TPA) events [53]. Because of the high peak laser intensity and coherency, DUV excimer laser irradiation can cause optically-induced transition and ionization in SiO_2 by TPA process, even when the photon energies of the DUV excimer lasers are less than the band-gap energy of SiO_2. This concept is covered in greater detail in subsequent sections.

10.1 Overview of UV Laser Induced Effects in SiO_2

The optical and mechanical properties of a material can be altered as a result of its exposure to energetic radiation or prolonged exposure. The form of alteration is called radiation damage, and it depends on the type of energetic beam, the acceleration potential of the beam, the composition of the irradiated material (e.g., whether the material is crystalline or glassy), and others [54]. In SiO_2, several manifestations of radiation damage can result from DUV excimer laser irradiation. The first is the induced absorption of SiO_2, related to the formation of color centers (e.g., the E' center and NBOHC). The second effect is material densification (i.e., density change), which involves structural reorganization and the alteration of the chemical composition in the material. The third manifestation is a photorefractive effect due to the formation of hydroxyl groups (SiOH), which causes changes in the optical index of refraction of the SiO_2 material [55]. As well, changes in the index of refraction can be derived from UV-induced absorption and densification.

These UV laser induced effects in SiO_2 are critical even at low laser fluences for lithography applications ($<0.1\,mJ/cm^2$). It has been proposed that the photon-induced breaking and forming of molecular bonds is the root of all damages, that are induced by DUV excimer laser irradiation [55]. As molecular bonds change, so do the structural arrangement and the chemical composition of the SiO_2. In the following, the three primary laser induced effects in SiO_2 are examined in greater depth.

10.1.1 Color Center Formation and Induced Absorption

The DUV excimer laser irradiation of SiO_2 gives rise to the formation of color centers. The two commonly encountered UV-induced color centers are the E' center ($\equiv Si^\bullet$) and the NBOHC ($\equiv Si-O^\bullet$).[2] Since the E' center has an absorption band centered at 5.8 eV, and the NBOHC has absorption bands peaking at 4.8 eV and 6.4 eV, SiO_2 becomes absorbing at these UV photon energies. Furthermore, the presence of these defect centers can inflict structural disorder on the SiO_2 network, which has a predominant influence on the photodegradation and stability of SiO_2 in the DUV regime [53].

Typically, the concentration of color centers increases with continuous DUV laser exposure; thus a gradual absorption increase (or transmission loss) in SiO_2, as a function of the DUV exposure, is commonly observed. The formation of color centers requires the presence of some precursors in the material; but when all the precursors are consumed, no new color centers are created [56]. This implies that the induced absorption increases initially, but eventually reaches a steady state. The number of laser pulses required to reach a steady state and the final absorption level are dependent on the exposure fluence and material composition. In addition to the degraded optical transmission, UV-induced absorption can provoke thermal effects (e.g., heating) in the SiO_2 that prompt further changes in the refractive index and structural deformation of the SiO_2.

The sequence of reactions responsible for the induced absorption in SiO_2 during a long-term DUV laser exposure is as follows. First, the absorbed laser radiation leads to the formation of excitons in the SiO_2. Although most of these excitons simply decay (i.e., returning to their ground state), others become trapped in localized states called precursors [56]. Each precursor (i.e., trapped exciton) decays and/or dissociates to form E' center and NBOHC. These color centers then react with the molecular hydrogen in the SiO_2 to form SiH and SiOH bonds. But, the resulting SiH and SiOH are convertible back to color centers and hydrogen by photolysis. For instance, the photolysis of SiH creates E' centers [56].[3] During the exposure period, an equilibrium develops between the two types of color centers (the E' center and NBOHC) on the one side, and the SiH and SiOH on the other. After a sufficiently long exposure, all the precursors that are initially present in the SiO_2 react to form color centers, and the induced absorption levels off to a steady state [56]. The relative number of color centers, and thus, the extent of the DUV absorption, is fluence-dependent since the equilibrium reactions involve photolysis.

The impregnation of a hydrogen molecule in the SiO_2 provides an effective method to suppress the formation of color centers, caused by UV radiation [57]. This is due to the fact that molecular hydrogen can react with the E' centers, and the E' centers are consumed; as result, the formation of

---
[2] The formation reactions of these color centers were described in Sect. 5.2.
[3] Refer to Table 5.3 for the associated defect formation reactions.

E' centers is suppressed and the induced absorption of the E' center band is reduced. Therefore, for SiO_2 material that contains hydrogen molecules, the laser-induced color centers will decay gradually after the termination of the irradiation.

Owing to the formation of color centers, the SiO_2 absorption changes with the length of the DUV exposure. An increased material absorption can be detrimental to the performance of microlithography equipment for several reasons. First, due to the higher absorption within the SiO_2 lens, more pulses are required to achieve the intended radiation dose on the wafers (reducing the system throughput). Secondly, the absorbed light heats the SiO_2 lens itself so that both its shape and index of refraction are altered (reducing the imaging quality).

These concerns are also relevant to CCD sensors. If the optical absorption of the SiO_2 layer increases with irradiation, it becomes more difficult for the sensor to detect the incident radiation, as fewer photons can reach the active Si layer to produce a useful signal. An increased absorption also leads to thermal effects that further degrade the optical properties of the SiO_2 layer and are detrimental to the CCD performance.[4]

10.1.2 Density Change

When SiO_2 is exposed to DUV laser radiation, two types of density changes in SiO_2 can occur, depending on the exposure conditions (e.g., the fluence and pulse length) and the material composition. The first effect, known as densification (or compaction), is an increase in the material density. The second effect is expansion (or rarefaction) which signifies a decrease in the material density. Compaction and expansion can occur simultaneously in an exposed piece of SiO_2 glass, but the dominance of one over the other depends on the exposure intensity and material composition.

Compared with compaction, expansion is significant only at very low fluences. The expansion effect is perceived to be due to the radiation-induced formation of β-hydroxyl (SiOH), where β-hydroxyl decreases the material density (i.e., expansion) and changes the chemical composition of the material [55]. The formation process of β-hydroxyl requires hydrogen. Therefore, aside from the laser fluence, the hydrogen content is another key factor that determines the severity of the UV-induced expansion in SiO_2.

Compaction or densification in SiO_2 is related to the optically-induced breaking or weakening of bonds that permits structural rearrangements and relaxations. UV laser induced densification in SiO_2 not only leads to an increase in the material density, but also triggers additional modifications in the material properties, including changes in the index of refraction (wavefront distortion), an induced surface deformation, and stress-induced birefringence [55]. The extent of the UV-induced density changes is dependent

---

[4] These issues are studied in Chap. 12.

on the sample, and the laser beam geometry. This is because the finite geometry of the laser beam irradiates only a finite area of the sample, thus, the unexposed region surrounding the irradiated region of the sample constrains the material's ability to shrink or expand on irradiation.

10.1.3 Photorefractive Effect

The photorefractive effect is a laser induced change of the index of refraction, associated with the hydroxyl (SiOH) formation in SiO_2. This effect complements the index change in the material that is caused by densification; therefore, the measured wavefront distortion is the sum of the wavefront distortions from the density change and the photorefractive effect. Although both the UV-induced density change and photorefractive effect can influence the index of refraction in the material, these effects must be treated separately for three reasons. First, the photorefractive effect affects the wavefront measurements but not the birefringence measurements. Secondly, the photorefractive effect is not subject to the constraints of the surrounding materials. Lastly, the adjustments in the index of refraction, due to the photorefractive effect, scale differently with the wavelength than the adjustments due to the density changes [55].

10.2 Active Defects in DUV

UV-induced absorption and defect formation in SiO_2 are primarily related to structural disorders in the network. The structural disorder in SiO_2 can be categorized as chemical disorder or physical disorder. Each of these disorders produces different color centers and has different ramifications on the radiation sensitivity of the SiO_2 in DUV.

Chemical disorder plays an important role in the optical absorption and radiation sensitivity of the material. The most common chemical disorder in the SiO_2 network is the point defect. Point defects include Si or O vacancies and their interstitials, homobonds, or over- or under-coordinated Si or O atoms. The chemical disorder in SiO_2 is suppressed by reducing the number of Si-Si bonds and SiOH groups, which contribute intense VUV absorption bands that peak at $7.6\,eV$ and $>7.4\,eV$, respectively. Also, the concentration of H_2 molecules, embedded in the SiO_2, must be low to avoid the restoration of the induced defects by the chemical reactions with H_2 [58]. For example, hydrogen reacts with SiH to generate more E' centers, and thus intensifies the chemical disorder.

Physical disorder is typically linked to the wide distributions of the Si-O-Si bond angle [∠(Si-O-Si)] and the size of the $(Si-O)_n$ ring structure. An example of physical disorder is the strained Si-O-Si bond; it introduces a wider bond angle distribution and smaller ring structures to the network. The physical disorder of the network structure controls the VUV absorption edge and the formation processes of the DUV-induced defects. Hosono et al. have reported that the VUV absorption edge shifts to a longer wavelength (i.e., a smaller optical band-gap) with the increased concentration of the strained Si-O-Si bonds. In addition, the direct dissociation of the strained Si-O-Si bonds is recognized to be the key defect formation channel with DUV excimer laser irradiation [58].

In most cases, the structural disorders in SiO_2 generate absorption bands in the UV-VUV region. Some of the principal defects and their corresponding absorption region are listed in Table 10.1. Each type of defect is unique to the preparation method and radiation history of the SiO_2 [59]. The defects which play a more active role with DUV excimer laser exposure are oxygen-deficient centers (ODCs), E' centers, and NBOHCs.

Table 10.1. The principal defects in SiO_2 and their optical absorption bands in the UV-VUV [59]

Defect in SiO_2	Absorption Band(s)
Hydroxyl group (OH)	7.0–8.2 eV
Peroxy linkage (POL)	6.5–7.8 eV
Oxygen-deficient center (ODC)	7.6 eV [ODC(I)] and 5.0 eV [ODC(II)]
E' center	5.8 eV
Non-bridging oxygen-hole center (NBOHC)	Centered at 4.8 eV and 6.4 eV

10.2.1 Oxygen-Deficient Centers (ODCs)

Oxygen-deficient centers (ODCs) in SiO_2 are classified as two variants: ODC(I) and ODC(II). The structural origin of the ODC(I) is the ≡Si–Si≡ homobond with an absorption band at 7.6 eV. ODC(II) is assigned to an unrelaxed neutral oxygen vacancy which is less stable than the ≡Si–Si≡ bonds. ODC(II) is characterized by an absorption band at 5.0 eV and a luminescence band at 4.3 eV. The structural configurations of these two types of ODCs are reflected in Fig. 10.1.

ODC(I) and ODC(II) respond differently to DUV laser irradiation. As shown in Fig. 10.2, the intensity of the ODC(II) (5.0 eV) band decreases, whereas the intensity of the ODC(I) (7.6 eV) band is constant, with the ArF (193 nm) excimer laser irradiation. Also, the reduction of the ODC(II) is accompanied by an increase in the E' center concentration; this suggests that the ODC(II) is consumed and converted to an E' center during the irradiation period.

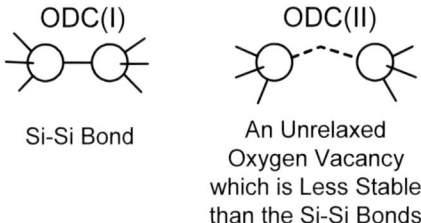

Fig. 10.1. Structural configurations for the two types of oxygen-deficient centers (ODCs) in SiO$_2$. The circle and line represent a Si atom and a bond to an oxygen atom, respectively (adapted from Imai et al. [59])

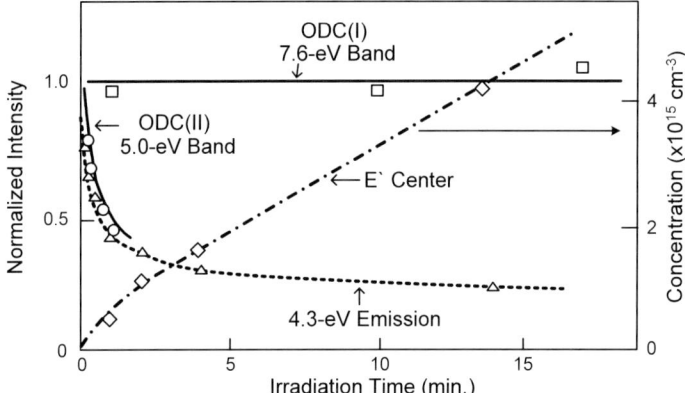

Fig. 10.2. Effect of ArF laser irradiation on the 7.6 eV and 5.0 eV absorption bands, the emission at 4.3 eV from the excitation of the 5.0 eV band, and the concentration of E' centers in SiO$_2$ glasses. The laser is operated at 30 Hz and 30 mJ/pulse (adapted from Imai et al. [59])

10.2.2 E' Centers

An E' center is an under-coordinated Si atom ($\equiv$ Si$^\bullet$) with an absorption band at 5.8 eV. The E' center and the NBOHC are the most frequently encountered color centers in SiO$_2$; they are formed by the photolysis of strained Si-O-Si bonds during DUV irradiation (which is expressed as (5.10))

$$(\equiv \mathrm{Si} - \mathrm{O} - \mathrm{Si} \equiv)^* \longrightarrow \ \equiv \mathrm{Si}^\bullet + \ ^\bullet\mathrm{O} - \mathrm{Si} \equiv , \tag{10.1}$$

where * denotes a strained bond. The E' center is also generated from an ODC with DUV excitation via the reaction (given in (5.6))

$$\equiv \mathrm{Si} - \mathrm{Si} \equiv \ \longrightarrow \ \equiv \mathrm{Si}^\bullet + \ ^+\mathrm{Si} \equiv + \ \mathrm{e}^- . \tag{10.2}$$

Usually, an E' center is created at the expense of the ODC(II), but can also be derived from the ODC(I). This is evident in Fig. 10.2, where the concentration of E' center increases as the intensity of the ODC(II) (5.0 eV) band decreases during the ArF laser irradiation. However, the concentration of E' center continues to increase even after the intensity of the 5.0 eV band becomes constant. This implies that the E' center is also derived from ODC(I). Nevertheless, the formation efficiency of the E' center from the ODC(II) is expected to be much higher than that from the ODC(I) [59].

If SiO_2 is subjected to heat treatment in an H_2 atmosphere during the fabrication process, ODCs ($\equiv$Si–Si$\equiv$) are converted into Si-H bond groups. Such groups become the dominant precursors of the E' centers in H_2-treated SiO_2 glasses. Imai et al. have suggested that the formation efficiency of E' centers in SiO_2 containing Si-H bonds is about two orders of magnitude higher than that in SiO_2 containing ODCs. Moreover, they observed some radiation-induced E' centers in H_2-treated glasses decay at room temperature (i.e., the concentration of E' centers drops after the DUV irradiation), which is attributed to the recombination of the E' centers with the radiolytic H_2 molecules [59]. The generation of E' centers during DUV exposure, along with the decay in the E' centers after irradiation, causes fluctuations in the induced absorption of SiO_2 and provokes instability in the DUV performance of CCD sensors. Evidence of the fluctuations in the absorption of SiO_2 due to DUV irradiation is provided in Sect. 10.3.4.

10.2.3 Non-Bridging Oxygen Hole Centers (NBOHCs)

Another active defect in DUV is an oxygen dangling bond known as a NBOHC ($\equiv$Si-O$^\bullet$). The NBOHC has two major absorption bands in UV, with peaks at 4.8 eV and 6.4 eV. The photoexcitation of the NBOHC at its 4.8 eV band emits a red photoluminescence (PL), centered at 1.9 eV. Two major mechanisms contribute to the formation of NBOHC in SiO_2 during the DUV laser irradiation. The first is the the cleavage of the strained Si-O-Si bonds to form a NBOHC and E' center pair, as given in (10.1). The second mechanism is the photolysis of SiOH groups, which was described as[5]

$$\equiv Si - OH \longrightarrow \ \equiv Si - O^\bullet + {}^\bullet H \ . \tag{10.3}$$

According to the studies by Kajihara et al., the SiOH concentration determines which NBOHC formation mechanism is dominant [60]. In wet SiO_2 where the SiOH concentration is high, the SiOH bond is efficiently photolyzed by F_2 laser photons to form a NBOHC. However, the recombination of the NBOHC with the dissociated hydrogenous species suppresses the build-up of NBOHC. For dry SiO_2 with a low SiOH concentration, the NBOHC formation by the dissociation of the strained Si-O-Si bonds is dominant; however, this process is inefficient [60]. Often, the NBOHC accumulates in relation to the number of F_2 laser pulses in dry SiO_2; this is attributable to the negligibly slow reverse (recombination) reaction of (10.3).

---

[5]Same as (5.13).

Strained Si-O-Si bonds are inherent in the a-SiO$_2$ network, and introduce band tail states near the fundamental absorption edge of the SiO$_2$. Thus, the color center formation by the photolysis of strained Si-O-Si bonds is inexorable in SiO$_2$ under DUV excitation. These UV-induced color centers contribute predominantly to the induced absorption of SiO$_2$, but they can also alter the electrical properties of the material. In the next sections, the specific laser induced effects in SiO$_2$ that occur with DUV excimer laser irradiation at the wavelengths of 248 nm, 193 nm, and 157 nm will be examined.

10.3 KrF and ArF Laser Induced Effects in SiO$_2$

Although similar defects are created in SiO$_2$ at the various DUV wavelengths, the photoinduced processes or reactions differ due to the different photon energies. For example, the defect formation in SiO$_2$ by KrF ($\lambda = 248$ nm, $E_{\mathrm{ph}} = 5.0$ eV) and ArF ($\lambda = 193$ nm, $E_{\mathrm{ph}} = 6.4$ eV) laser irradiation predominantly proceeds by a two-photon absorption process (since the two-photon energies at these wavelengths are greater than the band-gap of SiO$_2$). In the case of F$_2$ laser irradiation ($\lambda = 157$ nm, $E_{\mathrm{ph}} = 7.9$ eV), both one-photon and two-photon processes can contribute to the defect formation, depending on the F$_2$ laser power. An investigation by Kajihara et al. has indicated that irradiating with a high-power F$_2$ laser, above the threshold of 10 mJ/pulse/cm^2, creates the E' center by a two-step absorption process. Below the threshold, the creation of the E' center and NBOHC proceeds by a one-photon absorption process [53]. More discussion on the F$_2$ laser induced effects in SiO$_2$ is provided in Sect. 10.4. This section focuses on the various aspects of color center formation and the induced absorption in SiO$_2$ due to KrF and ArF excimer laser irradiation.

10.3.1 Wavelength vs. Rate of Color Center Formation

Since the photon energy of ArF and KrF lasers is lower than the band-gap energy of SiO$_2$, exciting an electron from the valence band to the conduction band is generally not accomplished by a single-photon absorption process at these wavelengths. Instead, a multi-photon absorption process must be considered, owing to the high-photon density and coherency of the pulsed DUV excimer lasers. Researchers have provided solid evidence that a two-photon absorption process is accountable for the defect creation in SiO$_2$ on ArF and KrF excimer laser irradiation. Arajo et al. have reported that the concentration of laser-induced defects increases as a function of the laser power squared in dehydrated SiO$_2$ [56]; this squared dependence is an indication of the two-photon process. A dependence of the defect formation rate on the photon energy is also observed, where the ArF laser (6.4 eV) generates the E' center more efficiently than the KrF laser (5.0 eV), whereas the XeCl laser (4.0 eV) generates no E' centers at all [53].

Arai et al. have studied the wavelength-dependence of the E' center generation in SiO_2 by measuring the spin density of the E' center as a function of the irradiating time with ArF and KrF excimer lasers and low-pressure Hg lamp. Fig. 10.3 shows that the largest E' center density is observed for ArF (6.4 eV) laser irradiation compared to the density generated by KrF (5.0 eV) laser; the low pressure Hg lamp yields a comparatively smaller density of E' centers [61]. Thus, the density of the E' centers depends on the wavelength of the radiation: higher photon energy induces a larger concentration of color centers.

In addition, Fig. 10.3 reveals that the E' center density increases linearly with KrF laser irradiation time, whereas the increase of the density of the E' center from ArF laser irradiation is slightly sub-linear [61]. Therefore, the defect formation efficiency and the formation processes in SiO_2 are directly influenced by the photon energy (or wavelength) of the radiation source.

10.3.2 Induced Absorption due to KrF Excimer Laser Radiation

In conjunction with the formation of E' centers, KrF (248 nm, 5.0 eV) excimer laser irradiation induces an absorption band in SiO_2 with a peak at 260 nm (4.77 eV) which is attributed to a NBOHC. Since the NBOHC absorption band is in the vicinity of the KrF laser photon energy, the NBOHC band is a major contributor to the induced absorption at 248 nm. Although E' centers are also generated by the KrF laser radiation, the E' center absorption band

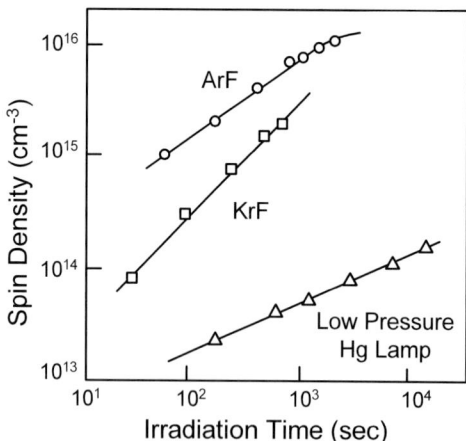

Fig. 10.3. The density of the E' center in SiO_2, induced by ArF, KrF excimer lasers and a low-pressure Hg lamp, as a function of the irradiation time. The ArF laser is operated with 45 mJ/pulse/cm^2 at 30 Hz. and KrF laser with 180 mJ/pulse/cm^2. The irradiation intensity of the 110 W Hg lamp at the sample is approximately 6 mW/cm^2 at 184.9 nm and 27 mW/cm^2 at 253.7 nm (adpated from Arai et al. [61])

at 5.8 eV does not significantly interfere with the transmission of SiO$_2$ at 248 nm. The KrF excimer laser radiation causes the density of the NBOHC to increase as the UV dose cumulates. This effect, in turn, leads to an increased absorption at 248 nm in SiO$_2$. The formation reactions of the NBOHC were described in Sect. 10.2.3.

10.3.3 Induced Absorption due to ArF Excimer Laser Radiation

The ArF (193 nm, 6.4 eV) excimer laser irradiation of SiO$_2$ gives rise to DUV-induced absorption that exhibits a distinct peak at 215 nm (5.8 eV), ascribed to the presence of E' centers. Also, NBOHCs are generated by ArF laser irradiation to provoke an absorption band close to 260 nm (4.77 eV), but compared to that of the 5.8 eV band, the absorption due to the NBOHCs has a weaker impact on the induced absorption intensity at 193 nm.

Section 10.2 indicated that the ODC is one of the main precursors of the E' center in UV-irradiated SiO$_2$. This argument is reinforced by the data in Fig. 10.4 for the SiO$_2$ absorption spectra, induced by ArF laser irradiation. The 5.8 eV band of the E' center grows, while the 5.0 eV band of the ODC(II) decreases, with the ArF laser irradiation time. However, the 7.6 eV band of the ODC(I) remains unchanged [61]. As a result, the E' center is produced primarily at the expense of ODC(II) after irradiation by a DUV excimer laser.

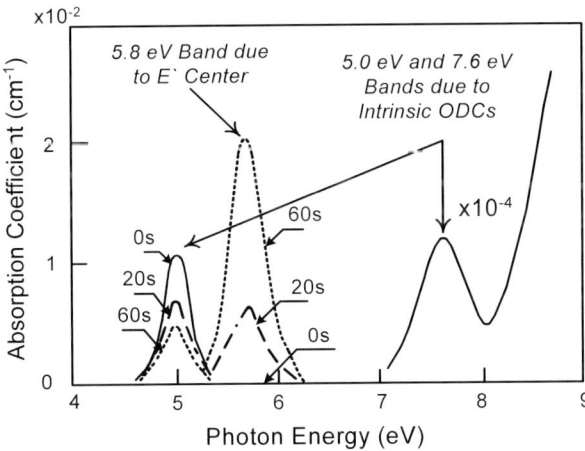

Fig. 10.4. The optical absorption characteristics of dehydrated synthetic SiO$_2$ glass. The growth of the 5.8 eV band due to the E' center, and the decrease of the 5.0 eV band due to the intrinsic ODCs are observed after irradiation by an ArF excimer laser of 45 mJ/pulse/cm^2 at 30 Hz. The irradiation times are indicated in the figure (adapted from Arai et al. [61])

10.3.4 Fluctuations in UV-Induced Absorption

Interestingly, the UV-induced optical absorption in SiO_2 can increase or decrease with the DUV exposure dose, depending on the irradiation conditions. A brief review of this phenomenon is provided for ArF excimer laser irradiation.

Changes in the optical absorption of lithographic-grade SiO_2 containing hydrogen have been examined by Ikuta et al. [57]. SiO_2 samples are irradiated intermittently by an ArF excimer laser (193 nm, 6.4 eV) under the operating mode of a lithography laser. The authors observed that the absorption intensity at 6.4 eV increases monotonically during the irradiation, and decreases gradually after the termination of the irradiation [57]. This fluctuating behavior in the induced absorption of SiO_2 is illustrated in Fig. 10.5. Figure 10.5(a) reflects the changes in the induced absorption intensity at 6.4 eV; here, the SiO_2 is irradiated intermittently by an ArF excimer laser with an energy density of $100\,mJ/pulse/cm^2$ for 0.5 Mpulses, and then the SiO_2 is left idle for 45 minutes at room temperature after the termination of the irradiation. These operations are repeated eight times. Figure 10.5(b) displays an enlarged view of the changes during the idle period.

As observed in Fig. 10.5, the absorption at 6.4 eV increases monotonically with a saturation tendency by the irradiation, but the absorption decreases gradually when the irradiation is interrupted. When the sample is re-irradiated, the induced absorption recovers to the same extent as before the termination of the irradiation after only a few thousands laser pulses. This fast re-darkening phenomenon (i.e., the rapid recovery to a higher absorption intensity) is described as a *rapid damage process*, because the rapid loss of optical transmission can significantly degrade the usefulness and lifetime of the SiO_2 material, when it is used as lens in photolithography systems [57]. It is notable that even though the sample is irradiated in an intermittent fashion, the overall absorption at 6.4 eV increases smoothly as the dose accumulates, which is similar to the net changes induced by a continuous UV irradiation.

Since the induced absorption at 6.4 eV is primarily attributed to the presence of E' centers (see Sect. 10.3.3), the fluctuations in the absorption intensity of SiO_2 are believed to be associated with the formation and restoration processes of the E' centers that were discussed in Sect. 10.1.1. After the E' centers are formed from the precursors by UV-induced reactions, the E' centers can react with the hydrogen to form SiH bonds. However, the photolysis of SiH bonds creates more E' centers. Thus, the concentration of E' centers can increase and decrease throughout the irradiation period. To specifically explain the fluctuation pattern in Fig. 10.5, Ikuta et al. have proposed the following processes for the photoinduced formation and restoration of defects in hydrogen-impregnated SiO_2 glass [57]. For the formation of E' centers, the primary reaction is the photolysis of the strained Si-O-Si bonds to create pairs of E' centers and NBOHCs; for the ArF excimer laser excitation,

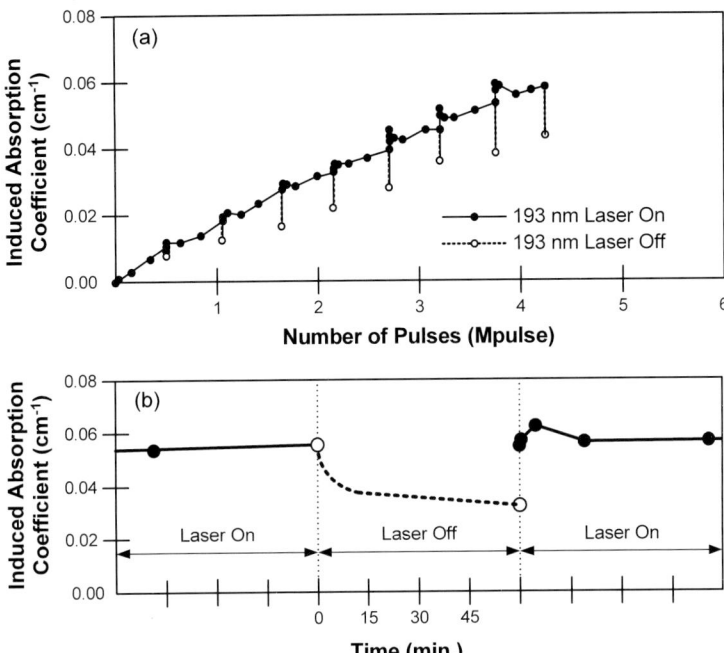

Fig. 10.5. (a) The changes in the induced absorption intensity at 6.4 eV, when SiO_2 is irradiated intermittently by ArF excimer laser with an energy density of 100 mJ/pulse/cm^2. The *closed* circles/*solid* lines are the absorption intensities during irradiation, obtained by an in-situ measurement; the *open* circles/*dotted* lines are the intensities after the termination of irradiation, obtained by an ex-situ measurement. (b) The enlarged changes in the induced absorption intensity at 6.4 eV after 3.8 Mpulses irradiation (adapted from Ikuta et al. [57])

this process occurs through a two-photon absorption process. ODCs are also induced by photoexcitation, which are further dissociated into E' centers. These two reactions are represented as follows:

$$\equiv Si\text{—}O\text{—}Si\equiv \text{ (strained)} \xrightarrow{ArF} \begin{cases} \overset{NBOHC}{\equiv Si\text{—}O^\bullet} + \overset{E^` Center}{{}^\bullet Si\equiv} \\ \overset{ODC}{\equiv Si\text{—}Si\equiv} + (1/2)\,O_2 \xrightarrow{ArF} \equiv Si^+ \; {}^\bullet Si\equiv \overset{E^` Center}{} + e^- \end{cases} \quad (10.4)$$

The induced E' centers and NBOHCs are consumed as they react with the hydrogen in the SiO_2 glass to form SiHs and SiOHs. The resulting SiHs are dissociated into E' centers and H atoms, presumably by a two-photon absorption process, by the ArF laser irradiation. In contrast, the SiOHs are minimally photolyzed, because they have no absorption at 6.4 eV (the photon energy of ArF laser). These reactions are summarized as

$$\overset{NBOHC}{\equiv Si-O^\bullet} + \overset{E'\,Center}{^\bullet Si\equiv} + H_2 \longrightarrow \equiv Si-H + HO-Si\equiv$$

$$2\,\overset{E'\,Center}{^\bullet Si\equiv} + H_2 \longrightarrow \equiv Si-H$$

$$\equiv Si-H \xrightarrow{ArF} \overset{E'\,Center}{^\bullet Si\equiv} + {}^\bullet H \qquad (10.5)$$

From this model, the three phases of the fluctuations in the induced absorption of SiO_2 are accounted for. In the first phase, an increase in the absorption intensity at 6.4 eV during irradiation is due to the formation of E' centers by two types of processes: a direct formation from strained Si-O-Si bonds and a dissociation of the photoinduced ODCs, given in (10.4). In the second phase, the absorption intensity decay process (or the fading of the induced absorption at 6.4 eV), after the termination of the irradiation, is attributed to the conversion of the E' centers into SiHs by the chemical reaction with the hydrogen in the SiO_2 (expressed in (10.5)). In the third phase, when the irradiation is resumed, the rapid damage process (or fast redarkening) is caused by the photolysis of the SiHs back to E' centers. In other words, SiHs act as precursors for the efficient photoformation of E' centers when the irradiation is continued after interruption; the formation of the E' center again contributes to the increase in induced absorption when the exposure resumes [57].

Since F_2 excimer laser photons (7.9 eV) have a higher energy than that of either KrF or ArF laser photons, the formation and restoration reactions of the E' centers are also accessible by F_2 laser photons. This suggests that fluctuations in the induced absorption are probable when SiO_2 is subjected to F_2 laser irradiation. Moreover, because F_2 laser photons are more energetic, the UV damage mechanisms in SiO_2 are expected to be of greater severity and complexity. The fluctuations in UV-induced absorption of SiO_2 have direct implications on the QE fluctuations in CCDs, when they are exposed to the F_2 excimer laser.[6]

10.4 F_2 Laser Induced Effects in SiO_2

In the previous discussion, it was concluded that the one-photon absorption process of KrF and ArF excimer laser irradiation has a negligible effect on

---

[6]More details are presented in Sect. 12.3.

the optically-induced transition in SiO_2 because the photon energies are lower than the band-gap energy of SiO_2. Thus, SiO_2 has a reasonable transmissivity at these wavelengths. However, defect formation is still inevitable, since the two-photon absorption process can be stimulated by the high photon density of the laser pulses. This is evident from the observation that the concentration of laser-induced defects increases as a function of the laser power squared.

For the case of F_2 excimer laser (157 nm, 7.9 eV), the photon energy is comparable to the fundamental absorption edge of SiO_2. As a result, the transparency of SiO_2 at 7.9 eV is often degraded due to optical absorptions, related to the structural imperfections such as Si-Si bonds, SiOH groups, and strained Si-O-Si bonds in the SiO_2 network. Chemical disorders such as Si-Si bonds and SiOH groups yield intense absorption bands, peaking at 7.6 eV and 7.4 eV, respectively. A one-photon excitation of these absorption bands by F_2 laser photon can lead to defect formation and the subsequent laser-induced effects. Kajihara et al. have reported a linear dependence between the concentration of laser-induced defects in SiO_2 and the F_2 laser power, which denotes a one-photon absorption process. Nevertheless, the two-photon absorption process still partake in the F_2 laser induced defect formation events, especially when SiO_2 is exposed to high-power F_2 laser irradiation [53].[7]

With the knowledge that Si-Si bonds and SiOH groups have absorption bands close to 7.9 eV that can hinder the optical transmission of SiO_2 during a F_2 laser exposure, it is essential to suppress these chemical disorders to yield a SiO_2 material with improved optical properties in DUV-VUV wavelengths. Also, the absence of these chemical disorders can provide a better environment for investigating the effects of physical disorder on F_2 laser induced damages in SiO_2. In addition, the concentration of H_2 molecules embedded in the SiO_2 samples needs to be low to avoid the restoration of the induced defects by the chemical reactions with H_2 (see reaction in (10.5)). For these reasons, researchers usually select point-defect-free SiO_2 samples for investigating F_2 laser induced effects in SiO_2; the outcomes from these investigations will be reviewed.

10.4.1 F_2 Laser Induced Defect Formation in SiO_2

The E' center and NBOHC are the two key defects created by F_2 laser irradiation in SiO_2, and are attributed to the dissociation of the strained Si-O-Si bonds; this reaction is given in (10.1). The strained Si-O-Si bond is the primary defect precursor for F_2 laser induced defects, and some of these strained bonds (three- or four-membered ring structures) in a SiO_2 network controls the optical absorption near 7.9 eV [58]. In Chap. 5, it was stated that the most stable and most populated bond angle, $\angle(Si\text{-}O\text{-}Si)$, is 145°, which belong to the six- to seven-membered ring structures. Three- and four-membered ring

---

[7] The dependence of the defect formation process on the F_2 laser power is addressed in Sect. 10.4.2.

structures introduce bond angles of 130.5° and 160.5°, respectively; these structures are heavily strained and relatively unstable compared to the rest of the network. Since the heavily strained bonds absorb near 7.9 eV, the F_2 laser photons can rupture these strained bonds so that they are converted into an E' center and a NBOHC. Consequently, the VUV absorption edge and the induced absorption at 7.9 eV is influenced by the concentration of the strained Si-O-Si bonds in SiO_2 [58].

An effective way to enhance the VUV transparency and suppress the F_2 laser induced defects in SiO_2 is by fluorine doping (F-doping). As the fluorine is incorporated into the SiO_2 network, Si-F bonds are formed. This reduces the fraction of the three- and four-membered ring structures (i.e., the strained Si-O-Si bonds) in the network, thus narrowing the distribution of the ∠(Si-O-Si), as illustrated in Fig. 10.6 [58]. Also, the F-doped SiO_2 shows a blueshift of the VUV absorption edge due to the higher absorption energy of the Si-F bonds and the reduction of the strained Si-O-Si bonds.[8] As a result of the F-doping, the SiO_2 undergoes an enhanced transmittance at 7.9 eV and an improved radiation resistance to F_2 excimer laser irradiation. This suggests that F-doped SiO_2 is suitable for F_2 excimer laser applications.

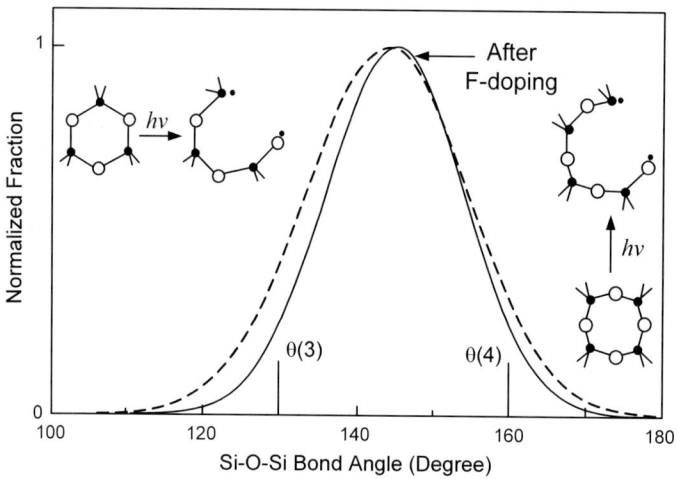

Fig. 10.6. A schematic model to explain the relation between the doping of fluorine, the degree of physical disorder, and the defect creation by F_2 laser irradiation in SiO_2 glass. When F-doping is introduced (*solid curve*), the distribution of the Si-O-Si bond angle becomes narrow and the peak position is slightly shifted to a large angle side (adapted from Hosono et al. [58])

---

[8] A *blueshift* is a decrease in the wavelength of radiation.

10.4.2 Dependence of Defect Formation on F_2 Laser Power

Depending on the F_2 laser intensity, the defect formation in SiO_2 proceeds either via a one-photon absorption process or a two-photon absorption process. Also, the concentration of the F_2 laser induced defects is a function of the laser power density and the pulse number. These dependencies between the defect formation process in SiO_2 and the F_2 laser irradiation parameters are discussed in this section.

10.4.2.1 Defect Formation vs. Pulse Number

The concentration of F_2 laser induced E' centers and NBOHCs increases with the number of laser pulses and the laser power density. Figure 10.7 displays the experimentally observed relationship between the induced absorption (or defect formation) and the number of accumulated F_2 laser pulses in SiO_2 [58]. The absorption intensities of the 4.8 eV band (NBOHC) and the 5.8 eV band (E' center) in SiO_2 samples increase, when the F_2 laser pulse number is increased, and approaches a certain saturation level, depending on the fictive temperature, T_f, of the specimen.[9]

According to Hosono et al., the data can be fitted by a first order kinetic reaction [58],

$$\Delta\alpha(0) = \Delta\alpha(\infty)\left[1 - \exp(k \cdot n)\right] , \qquad (10.6)$$

where $\Delta\alpha$ is the absorption coefficient of an induced optical band, n is the number of irradiated F_2 laser pulses, and k is the reaction rate constant. The value of k is independent of T_f, whereas $\Delta\alpha_{4.8eV}(\infty)$ or $\Delta\alpha_{5.8eV}(\infty)$ is larger at a higher T_f [58]. Thus, F_2 laser irradiation causes the absorption of SiO_2 at 4.8 eV and 5.8 eV to increase with the cumulative exposures, similar to the behavior that was reported for the KrF and ArF excimer laser irradiation in Sect. 10.3.

10.4.2.2 Defect Formation vs. Laser Power

The defect formation mechanisms in SiO_2 are dependent on the F_2 laser power. The changes in the optical absorption of a SiO_2 sample after F_2 laser irradiation at various laser intensities are depicted in Fig. 10.8 [53]. The absorption spectra indicate that the absorption of SiO_2 increases with the laser power. In addition, the E' centers and NBOHCs grow with increasing laser power, as suggested by the increase in $\Delta\alpha$ at the 5.8 eV band, and at the 4.8 and 6.4 eV bands, respectively. Although the E' center absorption band, centered at 5.8 eV, is negligible at a low F_2 laser power, increasing the laser power causes the E' center to grow faster than the NBOHC [53]. An

---
[9] The fictive temperature, T_f, is a value representative of the degree of the physical disorder of the Si-O-Si network. A higher T_f is generally associated to a higher disorder in the SiO_2 network and a VUV absorption edge at a lower energy.

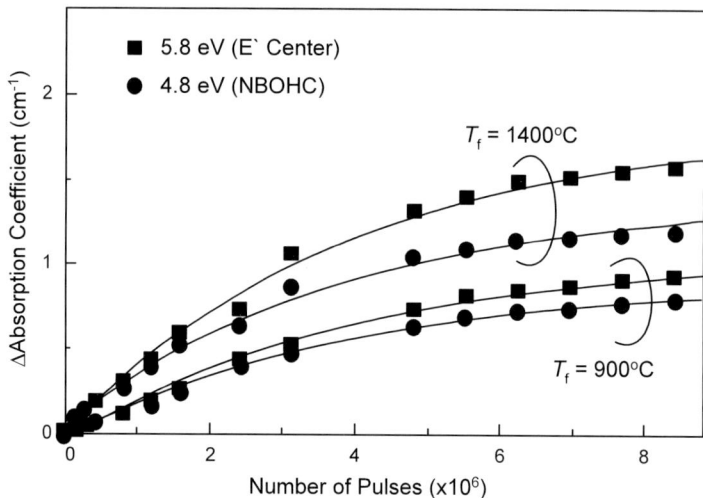

Fig. 10.7. The absorption intensities of the 5.8 eV and 4.8 eV bands as a function of the accumulative fluence of the F_2 laser for SiO_2 glasses (The SiO_2 samples have a very low SiOH and H_2 content. The F_2 laser is operated with a pulse width of 10 ns at 200 Hz, and a power density ~2.5 mJ/pulse/cm² at the sample position.) (Adapted from Hosono et al. [58].)

overall increase in the absorption of SiO_2 across the UV spectrum, for photon energies between 4.0 eV and 7.0 eV, is observed after the F_2 laser irradiation.

Peculiarly, F_2 laser irradiation causes a decrease in the absorption intensity for energy bands above 7.5 eV, as indicated by $\Delta\alpha < 0$ in Fig. 10.8. The reduced absorption in this region is believed to be due to the configurational change of the SiOH group from an isolated state to a hydrogen-bonded state, and the accompanying blueshift of the SiOH group's VUV absorption band. In other words, the isolated SiOH groups are optically-active at the F_2 laser wavelength, whereas the hydrogen-bonded SiOH groups are inert due to the blueshift of their absorption band beyond the F_2 laser wavelength [53]. This decreasing absorption is termed as *bleaching*; discussion of F_2 laser induced bleaching of the VUV absorption edge is deferred to Sect. 10.4.3.

A more detailed representation of the dependence of the defect concentrations on the F_2 laser power for SiO_2 is depicted in Fig. 10.9 [53]. It shows that the F_2 laser induced color center formation is more significant at higher laser intensities, which reinforces the data in Fig. 10.8. Also, Fig. 10.9 reveals that for F_2 laser intensities less than 10 mJ/pulse/cm², the induced concentrations of the E' centers and NBOHCs are proportional to the laser power density; this relation suggests that E'-NBOHC formation proceeds via one-photon absorption processes at these laser intensities. For the laser intensities above the threshold value of ~10 mJ/pulse/cm², the concentration of the E'

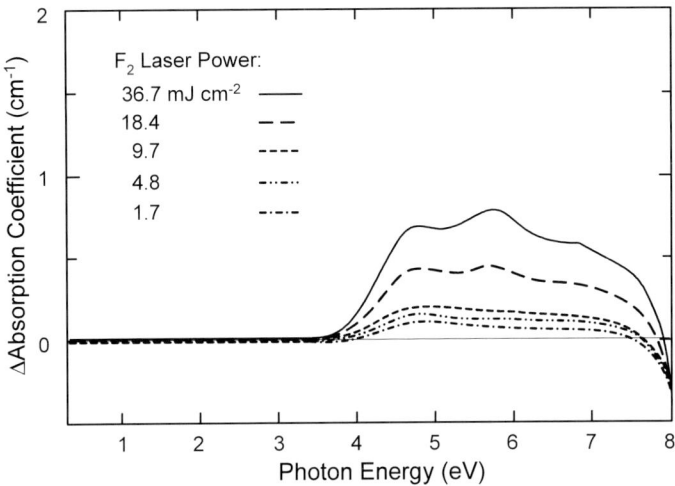

Fig. 10.8. Dependence of induced absorption of SiO_2 on F_2 laser power [F_2 laser is operated with pulse duration of 20 ns at 25 Hz, 3×10^4 pulses.] (Adapted from Kajihara et al. [53].)

centers, induced by the photolysis of Si-O-Si bonds, increases as a function of the F_2 laser power squared; this implies the formation process is correlated to a two-photon absorption process [53]. Another observation from Fig. 10.9 is the faster formation rate of E' centers compared to NBOHCs, as indicated by the larger slopes for the E' center.

The defect formation characteristics and their dependence on the F_2 laser power can be justified as follows. The dominant defect formation reactions, induced by the F_2 excimer laser in point-defect-free SiO_2 glasses[10], are the photolyses of both the strained Si-O-Si bonds and the SiOH groups; these reactions are expressed as

$$(\equiv Si - O - Si \equiv)^* \longrightarrow \equiv Si^{\bullet} + {}^{\bullet}O - Si \equiv \quad (10.7)$$

and

$$\equiv Si - OH \longrightarrow \equiv Si - O^{\bullet} + {}^{\bullet}H . \quad (10.8)$$

An E' center is created by the first reaction in (10.7), and both reactions contribute to the formation of a NBOHC. The quantum yield of the first reaction is very low ($\sim 10^{-4}$). In addition, the reverse reaction of (10.7) is insignificant at room temperature, because the recombination of the E' center and NBOHC is suppressed by the increased separation due to the strain energy, released during photolysis. Therefore, the concentration of E' centers

---

[10] The SiO_2 samples, used for F_2 laser studies, are point-defect-free SiO_2. Thus, the concentration of $\equiv$Si-Si$\equiv$ is too small to contribute to the defect formation of E' centers.

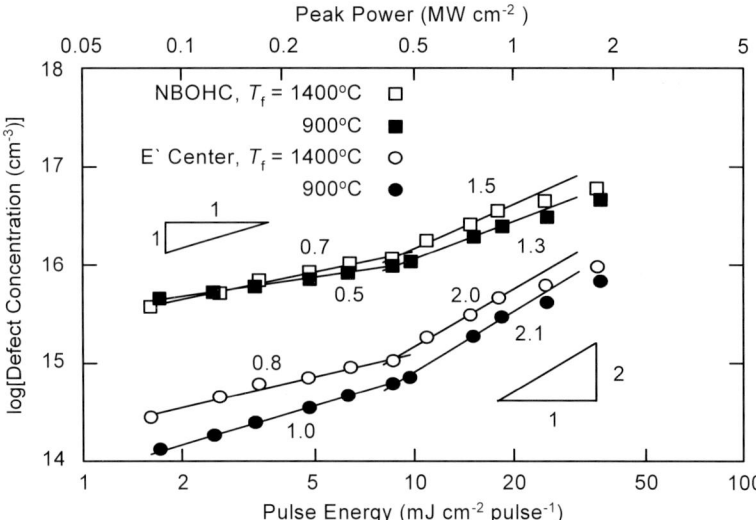

Fig. 10.9. The dependence of NBOHC and E' center concentrations on the F_2 laser power for SiO_2 (The F_2 laser is operated with a pulse width of 20 ns at 25 Hz, 3×10^4 pulses.) (Adapted from Kajihara et al. [53].)

increases linearly with the cumulative laser fluence and also increases with the concentration of strained Si-O-Si bonds. The saturation of the E' center concentration does not occur up to the fluence of 3×10^4 pulses, except at high pulse energies (>20 mJ/cm^2) [53].

On the contrary, the NBOHC is mostly the result of the reaction given in (10.8) rather than the first reaction, since the quantum yield of the second reaction is relatively large (∼0.1). However, the formation of NBOHCs by the photolysis of SiOH groups saturates rapidly (e.g., after a few hundred pulses). The early saturation of the NBOHC is presumably due to the exhaustion of the isolated SiOH groups that are available for photolysis and the rapid recombination of the NBOHC with the photoinduced hydrogen (the reverse reaction of (10.8)) [53]. Consequently, the concentration of NBOHCs, induced by F_2 laser irradiation, is larger than that of the E' centers and is comparably less sensitive to the pulse energy. The difference in the growth rate of the E' center and the NBOHC (i.e., the slope of the defect concentration vs. the pulse energy) in Fig. 10.9 is attributed to the distinction between their saturation behaviors. The slower growth rate (i.e., the smaller slope) of the NBOHCs is due to the more rapid saturation of the NBOHCs than that of the E' centers [53].

In summary, the concentration of UV-induced defects (i.e., E' centers and NBOHCs) increases with the F_2 pulse number and laser power. The defect formation in SiO_2 can proceed via a one-photon absorption process or a two-photon absorption process, depending on the laser power.

10.4.3 F_2 Laser Induced Bleaching of the VUV Absorption Edge

As seen in Fig. 10.8, F_2 laser irradiation can cause a reduction (or bleaching) of the optical absorption bands for energies above 7.5 eV. A similar behavior has been reported by Mizuguchi et al., where they have observed a decrease in the 157 nm absorption and the bleaching of the VUV absorption edge of SiO_2:OH glasses after F_2 laser irradiation [62]. Their results are summarized in Fig. 10.10, which shows an increase in the transmission at 157 nm from 62% to 71% by a F_2 laser irradiation. The F_2 laser induced bleaching of the VUV edge is depicted in the inset of Fig. 10.10, which shows an increase in the transmission, indicated by $\Delta\alpha < 0$, near 8 eV after F_2 laser irradiation. Thus, the near-VUV-edge absorption, at approximately 8 eV, is distinctly bleached by the F_2 laser irradiation [62]. The inset also reveals that weak absorption bands are induced at 5.8 eV (attributed to the E' centers) and at approximately 7 eV after the F_2 laser irradiation, as indicated by a slightly positive $\Delta\alpha$ at these energies.

The bleaching of the VUV absorption edge is attributed to two key processes: the transformation of the free SiOHs into hydrogen-bonded SiOHs, and the photolysis of the distorted Si-O bonds (e.g., the strained Si-O-Si

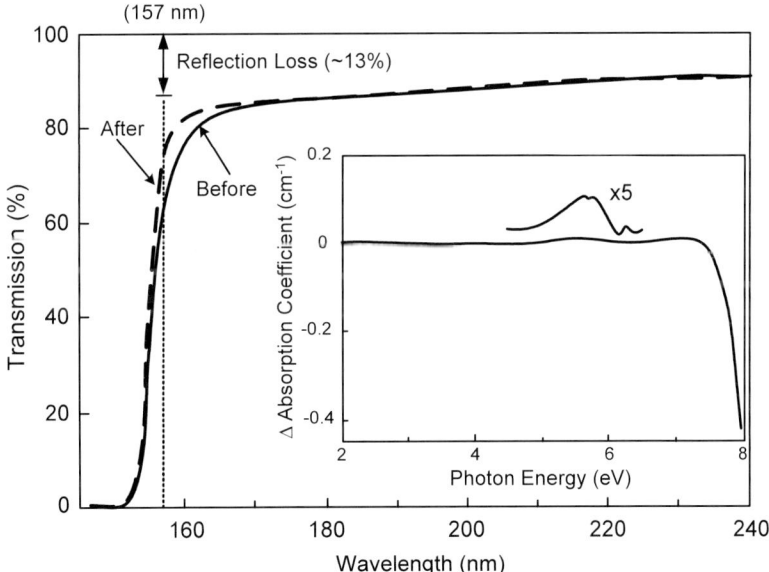

Fig. 10.10. The VUV–UV optical transmission spectra of SiO_2 before and after the F_2 laser irradiation. Inset: the F_2 laser-induced changes in the visible-VUV absorption spectrum (The F_2 laser is operated with 6 ns pulses at 400 Hz, power density of $\sim$10 mJ/pulse/cm^2 at the sample position, 3.6×10^5 pulses.) (Adapted from Mizuguchi et al. [62].)

bonds) [62]. The term, free SiOHs, denotes SiOHs that have no additional hydrogen bonds to the oxygen atoms. The free SiOH groups and the distorted Si-O bonds exhibit a significant absorption at or near 157 nm. Consequently, if the concentrations of these species are reduced (by photoinduced transformation and photolysis), the transmission at 157 nm increases theoretically.

To examine how F_2 laser irradiation interacts with the Si-O bonds and the SiOH groups to provoke photoinduced bleaching near the VUV absorption edge, the two important aspects that distinguish F_2 laser irradiation from KrF and ArF laser irradiation should be considered. First, the energy of the F_2 laser photons is just below the fundamental absorption edge of SiO_2 glass. Since the absorption edge is usually extended to lower energies by the exciton-phonon interaction (Urbach's tail in a-SiO_2) and the structural disorders in the a-SiO_2 network, any Si-O bond can be excited by a F_2 laser photon with some (small) probability. However, this probability is larger for Si-O bonds at the distorted sites with an electronic energy that is lower than that of normal bonds [62]. The band-to-band excitation of the distorted Si-O bonds by F_2 laser irradiation can stimulate a selective bleaching of the distorted Si-O sites in a photochemical reaction. One possibility is the hydrogenation of an electronically excited Si-O bond at the distorted sites such as in strained Si-O-Si bonds,

$$(\equiv Si - O - Si \equiv)^* \xrightarrow{h\upsilon(7.9eV)} \equiv Si - O - H + H - Si \equiv . \tag{10.9}$$

This reaction implies that the distorted Si-O bonds are consumed during the F_2 laser irradiation to form SiOH and SiH. As the concentration of the distorted Si-O bonds decreases, the optical absorption at 157 nm, due to these distorted bonds, diminishes.

The second unique feature of F_2 laser irradiation is that the 7.9 eV photons can directly excite the O-H bond in SiOH groups. Thus, F_2 laser exposure can cause the photolysis of SiOH, which subsequently generate oxygen radicals (e.g., NBOHCs) and highly mobile atomic hydrogen (see (10.3)). Since the free SiOH groups have an absorption band peaking near 7.4 eV, the consumption of these groups reduces the induced absorption at 157 nm (7.9 eV) and contributes to the bleaching of the VUV absorption edge [62].

In conclusion, during the F_2 laser irradiation, the distorted Si-O bonds and SiOH groups (that absorb at 157 nm and the VUV edge) in SiO_2 are depleted by their conversion to other species. Since these newly-generated species do not absorb at 157 nm or the VUV edge, consequently, there is an enhanced transmission of 157 nm photons and a photoinduced bleaching of the VUV absorption edge after F_2 excimer laser irradiation. Such a reduction in the optical absorption in SiO_2 near the VUV edge is important to the F_2 laser induced QE fluctuations in CCDs. If the absorption of SiO_2 decreases with the F_2 laser irradiation, more incident photons can reach the Si active layer of the CCD to generate signal charges. This results in an increase in the QE with the cumulative F_2 laser dose.[11]

10.5 UV-Induced Charging of SiO_2

The UV-laser induced effects, discussed in the preceding sections, was mainly focused on the optical properties of SiO_2. The electrical properties of SiO_2 are also susceptible to changes from DUV laser irradiation. A dominant effect is the UV-induced oxide charging, which can emerge from a combination of processes including ionization damage in the oxide, generation of charged defects, trapping of carriers by defects, and photoemission of electrons from Si to SiO_2. For instance, the oxide layer becomes negatively charged due to the UV photoemission of electrons from the Si into the SiO_2.[12] Also, the ionization damage can generate trapped holes to induce positive charging of the oxide layer.[13] The electrically-active defects in SiO_2, presented in Chap. 5, can contribute to the additional charging of SiO_2. The resulting charge accumulation in SiO_2 inflicts serious concerns on the electrical performance and stability of semiconductor devices.

The effect of UV-induced fixed oxide charge modifications in SiO_2 films has been investigated by Fiori et al. [63]. Various types and qualities of SiO_2 films (e.g., wet and dry SiO_2, with or without post-growth annealing), grown on MOS structures, are subjected to KrF (248 nm) excimer laser irradiation. Electrical characterization is performed to determine the influence of the DUV irradiation on the fixed oxide charge density, N_f. The changes in N_f as a function of the accumulated 248 nm dose for unannealed oxides are displayed in Fig. 10.11. For unannealed dry oxides, N_f decreases with increasing the 248 nm dose. For unannealed wet oxides, N_f decreases with the 248 nm dose, and becomes negative after a total dose of $\sim 20 \, \text{J/cm}^2$. However, after an accumulated dose of $\sim 2000 \, \text{J/cm}^2$, N_f increases for both types of unannealed oxides, and is believed to correspond to the onset of defect creation and oxide degradation [63].

Interestingly, the postgrowth-annealed oxides undergo a different kind of modification in N_f as a function of the accumulated 248 nm dose compared to unannealed samples, as illustrated in Fig. 10.12. For annealed oxides, N_f increases with the accumulated UV dose, and the rate of increase in N_f is more prominent in dry SiO_2 than in wet SiO_2 [63]. This suggests that the oxide charged defects (e.g., oxygen vacancy centers) are more readily created in dry SiO_2 than in wet SiO_2 during UV irradiation, and contribute to the increase in N_f.

---

[11] Section 12.3 provides a more detailed analysis on the DUV damage of CCDs.
[12] UV photoemission was discussed in Sect. 8.2.1 and Chap. 9.
[13] The characteristics of the radiation-induced oxide-trapped charges were presented in Chaps. 7 and 8.

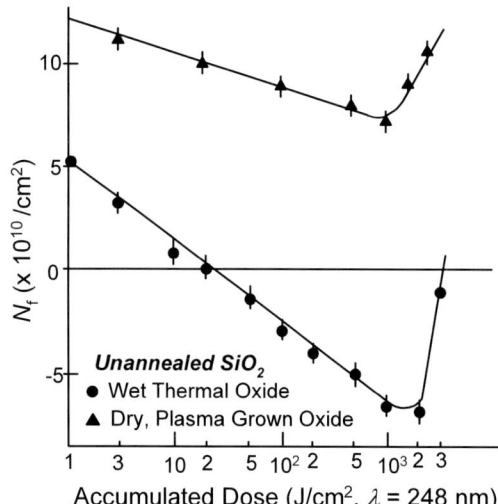

Fig. 10.11. The measured fixed oxide charge density, N_f, in unannealed oxides as a function of the accumulated dose of 248 nm irradiation (KrF excimer laser is operated with a pulse width of 20 ns, peak pulse power <300 mJ/cm^2 at the sample surface, and repetition rate <3 Hz to avoid sample heating.) (Adapted from Fiori et al. [63].)

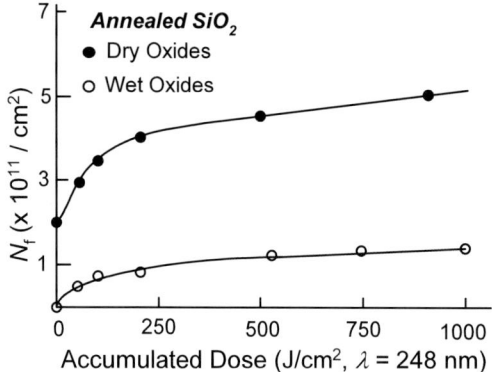

Fig. 10.12. The measured fixed oxide charge density, N_f, in annealed oxides as a function of the accumulated dose at 248 nm. The wet and dry thermal oxides are annealed at 950°C for 30 min. in N_2 after growth (adapted from Fiori et al. [63])

Understanding the different behaviors of the UV-induced oxide charge densities in Fig. 10.11 and Fig. 10.12 requires insight into the nature of annealed and unannealed oxides. In the case of unannealed oxides, when electrons are photoinjected into the SiO$_2$ from the Si substrate or excited from the SiO$_2$ valence band edge, the electrons can become trapped at dangling

10.5 UV-Induced Charging of SiO$_2$

hydroxyl linkages to result in hydrogen, H^0, emission and the formation of negatively charged non-bonded oxygens [63],

$$\equiv \text{Si} - \text{OH} + e^- \longrightarrow \equiv \text{Si} - \text{O}^- + \text{H}^0 . \quad (10.10)$$

The generation of negatively charged non-bonded oxygens causes a net decrease in the positive oxide charge density in unannealed oxides; this process is responsible for the decrease in N_f with the 248 nm dose, observed in Fig. 10.11. Since unannealed wet oxides contain more SiOH groups than dry oxides, the reaction in (10.10) is more favored in wet oxides, leading to a larger variation of N_f for wet oxides. Also, this implies that a larger quantity of negatively charged species ($\equiv$Si-O$^-$) is generated in wet unannealed oxides, resulting in negative N_f values as the UV dose accumulates.

Alternatively, the reduction in N_f, observed in unannealed oxides, originates from the annihilation of the positive fixed oxide charges. This process occurs by the trapping of electrons, excited from the SiO$_2$ valence band edge or photoinjected from Si into SiO$_2$, to annihilate the positive oxide charge and the E' center, which is expressed as:

$$\equiv \text{Si}^+ \bullet \text{Si} \equiv + e^- \longrightarrow \equiv \text{Si} - \text{Si} \equiv + h\upsilon . \quad (10.11)$$

As a result, the positive oxide charges (in the form of oxygen vacancy centers) are reduced by the 248 nm irradiation. This is the dominant reaction that is responsible for the N_f reduction in dry unannealed oxides.

Contrary to the annihilation of the positive charges in unannealed oxides, the creation of fixed oxide charges is observed in annealed oxides as perceived by the increase in N_f with the UV dose in Fig. 10.12. This characteristic suggests that in annealed wet oxides, the OH concentration is reduced by annealing (indicating that the reaction in (10.10) is not very favorable), and other species capable of producing positive fixed oxide charges are present. The data indicates that the UV-induced N_f in annealed oxides are nearly an order of magnitude larger than those in unannealed oxides so that annealing may have created large numbers of defect precursors in the SiO$_2$ network [63]. In annealed SiO$_2$, positive fixed oxide charges can result from the multiple step process, involving the recoupling of a positively charged dangling bond; namely,

$$\equiv \text{Si} - \text{O} - \text{O} - \text{Si} \equiv \longrightarrow \equiv \text{Si} - \text{O}^\bullet {}^\bullet \text{O} - \text{Si} , \quad (10.12)$$

$$\equiv \text{Si} - \text{O}^\bullet + h\upsilon \longrightarrow \equiv \text{Si} - \text{O}^+ + e^- , \quad (10.13)$$

and

$$\equiv \text{Si} - \text{O}^+ + \equiv \text{Si} - \text{O} - \text{Si} \equiv \longrightarrow \begin{array}{c} \equiv \text{Si} - \text{O} - \text{Si} \equiv \\ | \\ \text{O} \\ | \\ \text{Si} \\ ||| \end{array} . \quad (10.14)$$

Thus, according to (10.12) and (10.13), positively charged oxide defects ($\equiv$Si–O$^+$) are generated from peroxy bridges ($\equiv$Si–O–O–Si$\equiv$) to result in an increase in N_f. At the same time, the defects are consumed by the reaction expressed in (10.14). These two competing reactions are potential causes for the saturation behavior of N_f at the higher doses in Fig. 10.12. The process is expected to be particularly efficient in dry oxide, since high-temperature anneals tend to produce large numbers of peroxy bridges, that is consistent with the results in Fig. 10.12 [63].

Evidence of UV-induced changes in the oxide charge density has also been seen from the studies on the VUV and plasma treatment of thermal oxide on a Si substrate [64]. Usually, the thermally grown oxide contains positive charges within the order of 10^{10} to 10^{20} cm^{-2}, depending on the growth process. The bridging oxygen vacancy defect (O$_3\equiv$Si$^{\bullet}$ $^{\bullet}$Si$\equiv$O$_3$), most commonly the E' center, is a source of positive charges. A decrease of the positive oxide charges is observed on VUV exposure, and is suggested to be associated with the annihilation of E' centers or the generation of negative compensating charges by the VUV radiation [64].

It appears then that UV irradiation can induce changes in the electrical state of the oxide layer. N_f can increase or decrease with UV irradiation, depending on the type and processing conditions of the SiO$_2$ and the cumulative UV dose. These fluctuations in N_f definitely disrupt the electrical stability of CCDs with DUV irradiation.[14]

10.6 Summary of the UV-Induced Effects in SiO$_2$

Studies of the material interaction with UV radiation in the literature provide strong evidence that the UV damages, experienced by CCDs, are largely due to the UV-induced effects in the SiO$_2$ layer, and partly due to changes induced at the Si-SiO$_2$ interface. This section summarizes the various UV-induced effects in SiO$_2$, from both the optical and electrical perspectives, that are most relevant to the DUV-induced degradation of CCD sensors. These effects will provide a basis for analyzing the mechanisms responsible for such CCD degradation in 12.

10.6.1 Optical

Often, the DUV-VUV transmission of a-SiO$_2$ is affected by the absorption bands in the UV due to the presence of (pre-existing or UV-induced) point defects and impurities in the oxide network. Furthermore, the exposure to UV results in bond rearrangement and structural changes in SiO$_2$, such as

---

[14] An analysis of the effect of UV-induced oxide charging on the CCD sensor's characteristics is presented in Sect. 12.3.

densification or compaction. Some of the new species, triggered by the UV-induced changes in SiO_2, transform into point defects and color centers that introduce additional absorption bands in the UV region, further degrading the transparency of the SiO_2. Studies have shown that the DUV absorption of SiO_2 is directly linked to the strained Si-O-Si bonds [53]. Not only do the strained Si-O-Si bond absorb DUV-VUV photons, the photolyses of these strained bonds into E' centers and NBOHCs are the main defect formation mechanism during DUV irradiation of SiO_2. The SiO_2 absorption characteristic is controlled by the concentration of these color centers, which changes as a function of the cumulative DUV exposure. Ikuta et al. have observed that the UV-induced absorption intensity of SiO_2 at 6.4 eV increases with continuous ArF (193 nm) excimer laser exposure and decreases when the exposure is interrupted (Sect. 10.3.4) [57]. A similar fluctuation pattern is observed in the extrinsic QE of CCDs at 157 nm (Chap. 12). These temporal variations in the SiO_2 absorption are due to the formation and restoration of UV-induced color centers. Furthermore, the F_2 laser induced bleaching of the VUV absorption band and the decrease in 157 nm absorption of SiO_2, reported by Mizuguchi et al., are important to the investigation of CCD sensors at 157 nm [62]. These reactions perturb the optical performance of CCD sensors in DUV.

10.6.2 Electrical

Not only are the optical properties of SiO_2 altered by the formation of color centers and point defects, but also the electrical properties are affected by these UV-induced defects. Many of the point defects in SiO_2 can produce charged centers and influence the electrical behavior of Si-SiO_2 devices. Some examples of electrically-active defects in SiO_2 are briefly described:

- The elongated Si-O bond is a possible origin of the fixed oxide charges in the SiO_2 bulk or at the Si-SiO_2 interface.
- The strained bonds can cause a shift of the valence band edge and alter the effective band-gap energy.
- The breaking of strained Si-O bonds by radiation can generate e-h pairs, which contribute to the conduction processes or give rise to traps in the SiO_2.
- The re-creation of broken Si-O bonds leads to the relaxation of the surrounding lattice, creating localized energy levels in the SiO_2 band-gap.
- The E' center is a charged, relaxed, oxygen vacancy in SiO_2.
- The NBO defect behaves as a negatively charged point defect if the NBO accepts a bond electron; also, the NBO can stimulate a trap state in the band-gap.

These intrinsic point defects can be pre-existing (i.e., from the fabrication process) or induced by UV radiation. The presence of these defects and their interaction with the oxide network has a significant impact on the electrical characteristics of Si-SiO$_2$ structures.

UV radiation can induce charged oxide defects to further alter the electrical properties of the Si-SiO$_2$ interface. For instance, the electrons in Si can be photoemitted into the SiO$_2$ layer by the UV excitation, where they become trapped and charge the oxide negatively (Sect. 8.2.1). Other sources of UV-induced electrically-active defects in oxide include color centers behaving as charged species, oxide-trapped charges resulting from trapping of carriers that are photoinjected into the SiO$_2$, the generation of oxide charged defects, and ionization reactions.

The presence of UV-induced oxide charges can either increase or decrease the QE of CCDs, depending on the charges' polarity and concentration. If the SiO$_2$ layer acquires a net negative polarity from the UV-induced charges, the holes in the Si layer are attracted to the Si-SiO$_2$ interface to passivate the interface traps. The passivation of the interface traps reduces the electron trapping, so that an improvement in the QE is expected. Alternatively, ionizing radiation can induce a positive oxide charge by generating defects or hole trapping in the oxide; as a result, the trapping dynamics and electrical properties in the oxide and interface are modified in a different fashion (Sect. 7.2). Evidence of UV-induced modification of the oxide charge density as a function of the cumulative UV dose was discussed in Sect. 10.5. These dose-dependent fluctuations are potential origins of the DUV-induced instability of CCD sensors.

11 UV Laser Induced Effects at the Si-SiO$_2$ Interface

In Chap. 7 and Chap. 8, it was demonstrated that ionizing radiation can induce damages in the SiO$_2$ and Si-SiO$_2$ interface layers. UV radiation can induce defects and provoke physical changes (e.g., to induce or relieve the stress) at the Si-SiO$_2$ interface. This can impact the interface state density and the interface trapping and detrapping processes, which then alter the electrical behavior of semiconductor devices. The characteristics of radiation-induced interface states and their effect on the CCD performance (e.g., dark current and flatband voltage) were described in Sect. 8.2. Research groups have found other forms of interface changes that are induced by DUV excimer laser irradiation.

An example of these transformations is the UV laser induced structural modifications at the Si-SiO$_2$ interface, reported by Lu et al. [65]. They observed the phenomenon of laser-induced periodic surface structure (LIPSS) in a Si substrate, covered with a thin layer of SiO$_2$, on exposure to KrF excimer laser irradiation. Since SiO$_2$ is transparent to KrF (248 nm) laser photons, the laser-material interaction occurs mainly at the Si-SiO$_2$ interface near the Si substrate. A large part of the laser energy is transferred into heat on the Si side of the interface, owing to the lower thermal conductivity of SiO$_2$ than that of Si. Thus, an appropriate laser fluence can cause melting in the Si substrate and structural rearrangements at the interface, while the surface SiO$_2$ layer remains intact. Different thicknesses of the SiO$_2$ layer apply different surface tensions on the melted Si surface, and LIPSS with different periodicities can form. For thin SiO$_2$ layers, KrF excimer laser irradiation produce periodic microstructures at the Si-SiO$_2$ interface by a single pulse, if the laser fluence is large enough (e.g., 710 mJ/cm^2). When the SiO$_2$ layer is thick, more than one laser pulse is required to trigger a LIPSS. In this case, LIPSS in the form of circular patterns can be observed due to interface defects, and the periodicity of the ripple structure depends linearly on the SiO$_2$ thickness [65]. The formation of a LIPSS is a clear indication that bond structure modifications occur at the interface due to DUV laser irradiation. Electrically-active interface defects can emerge from these interfacial modifications, provoking changes in the electrical behavior of Si-SiO$_2$ devices.

There is another instance of UV-induced surface changes in a Si-SiO$_2$ structure. Kurosawa et al. have observed that argon (9.8 eV, 126.5 nm) excimer laser irradiation transforms the surfaces of the SiO_2 to Si (i.e., the 9.8 eV induces desorption and Si precipitation in the surface layers of the SiO_2) [66]. The maximum thickness of the Si precipitation layer correlates to the penetration depth of the argon laser light into the SiO_2 (i.e., ~50 nm). However, such a modification does not occur with krypton (8.5 eV, 145.9 nm) excimer laser irradiation. The laser output energies of argon and krypton lasers for this study are, respectively, approximately 10 and 25 mJ/pulse (i.e., energy fluences of 125 and 250 mJ/cm^2) on the sample's surface [66].

Since the 9.8 eV photons (from the argon laser) are larger than the fundamental band-gap energy of the SiO_2, the argon laser photons can induce the band-to-band transition of electrons in SiO_2 to generate excitons efficiently by a one-photon absorption process. A high density of excitons can stimulate breaking of Si-O bonds, which can result in Si precipitation and oxygen desorption on the SiO_2 surface. Thus, the VUV photons from the argon laser are capable of causing surface alteration [66]. When the argon excimer laser is operated at a higher intensity, more serious optical damage occurs in the surface layer of the synthetic quartz glass. For example, an irradiation damage pattern of 5 mm in diameter on the SiO_2 glass surface forms after only one laser shot with an energy density of 280 mJ/cm^2 and a pulse width of 5 ns [66]. This damage is ascribed to a photochemical effect, which results in the breaking of Si-O bonds and the subsequent isolation of crystalline Si in the surface layer of the SiO_2.

These investigations demonstrate that DUV lasers can induce bond breakages and structural rearrangements at the Si-SiO_2 interface and on the material surface. These interfacial changes can inflict serious impairment on the electrical performance of semiconductor devices.

In addition to the DUV-induced effects in Si, SiO_2, and the Si-SiO_2 interface considered in Chap. 4 to Chap. 11, numerous other effects can be provoked by DUV irradiation such as photoinduced oxidation, photoinduced oxygen loss in SiO_2 films, and UV-induced ablation. These reactions influence the optical and electrical characteristics of Si-based devices. However, these processes usually involve very high laser intensities and are less relevant for the purposes of the topic investigated in this book. In the next chapter, the DUV-induced effects, reviewed in the preceding chapters, are used to analyze and justify the CCD behavior at DUV wavelengths.

Part V

Interaction of DUV Radiation with CCD Sensors

12 CCD Measurements at 157 nm

Due to the growing demands of the lithography and semiconductor inspection industries for imaging at DUV wavelengths, DUV-sensitive CCD cameras that are fast, responsive, and stable are desirable. The challenges of utilizing CCDs in DUV and the currently available techniques for enhancing the UV sensitivity of CCDs were discussed in Chap. 3. In this chapter, a DUV-sensitive frontside-illuminated linear CCD image sensor design is introduced. Here, the overlying oxide of the photosensitive region is partially thinned to enhance the sensor's responsivity to DUV radiation. The key advantage of this design is that it is built by adapting the standard CCD process flow. This is in contrast to other popular UV-sensitive CCD sensors which often require costly and non-standard fabrication procedures (e.g., phosphor coating and backside thinning). Consequently, the concept of the thinned frontside-illuminated CCD sensor should offer a lower cost alternative for DUV imaging.

The DUV characteristics of the thinned frontside-illuminated linear CCD sensors are evaluated by exposing samples to a F_2 excimer laser at a wavelength of 157 nm. The QE and dark current are measured. The results are reported in this chapter, along with an analysis of the mechanisms that are responsible for the UV-induced damage in CCDs. The goal is to explain the CCD degradation effects in DUV by connecting the 157 nm experimental results with the findings that are gathered from the various research areas. These relevant topics were presented in the preceding chapters (Chap. 4 to Chap. 11), including a discussion of the properties of a Si-SiO_2 system, the interaction between the radiation and materials, the radiation damages in CCDs, and the UV-induced effects in SiO_2 for lithography applications.

12.1 Experiment Description

12.1.1 Laser Setup

For this experiment, the F_2 excimer laser (LPF200), manufactured by Lambda Physik, is used as the radiation source. The laser emits 157 nm photons and some output in the red. Neon gas is used to run the F_2 laser which reduces the red component of the laser output. Since the oxygen in the atmosphere absorbs 157 nm photons, the F_2 laser output is confined inside a vacuum chamber that has been purged with nitrogen (N_2) gas prior to the experiment.

The CCD sensors are exposed to a 157 nm F_2 excimer laser at a pulse frequency of 10 Hz to 100 Hz and a pulse width of 15 ns. The intensity of the laser output is controlled by a high voltage input to the laser. An attenuator, located at the entrance of the chamber, is capable of attenuating the raw laser output energy by 10% to 70%; the attenuator outputs an $8 \times 20\,mm^2$ rectangular laser beam at the exit port. The laser beam is then transmitted through a $4 \times 4\,mm^2$ aperture, which is positioned at the exit port of the attenuator.

A series of fused silica (SiO_2) plates are installed after the aperture to provide additional attenuation for the laser signal. Each 1 mm thick silica lens is rated for a 12% transmission approximately, and a 1.5 mm thick silica lens is rated for a 16% transmission approximately. Sufficient attenuation is necessary to ensure that the CCD sensors do not saturate during the response and QE measurements, and that the sensor surface is not physically damaged (e.g., ablation or etching) by the energetic laser beam. A phase mask filter is positioned after the fused silica plates and in front of the CCD sensor to divert the red component of the F_2 laser beam away from the CCD sensor; therefore, only the 157 nm photons of the laser beam are directed at the CCD sample. The phase mask transmits approximately 50% of the 157 nm laser energy to the CCD sensor. The energy of the laser output is measured by a Molectron (EPM 1000) Laser Energy Power Meter inside the chamber. The experiment configuration is depicted in Fig. 12.1. The room light is turned off to avoid interference on the CCD response.

The CCD sensors are set up in a windowless configuration and are operated by a DALSA Piranha camera. The response of the camera's video output signal, V_{OS}, is monitored throughout the experiment for measuring the CCD response and the dark current.

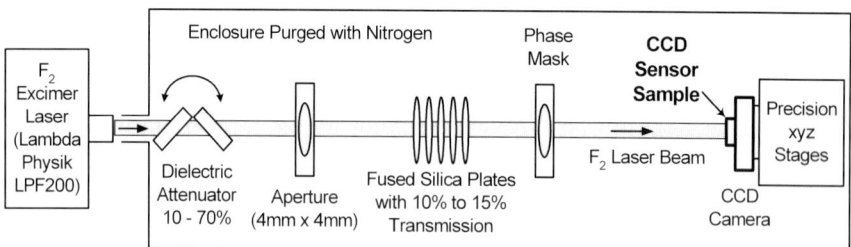

Fig. 12.1. The experimental setup for the 157 nm measurement

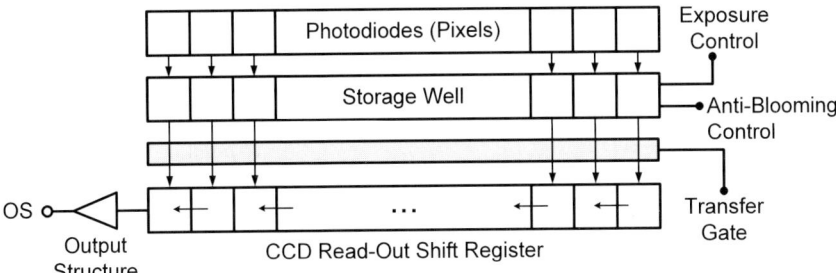

Fig. 12.2. A block diagram of the linear CCD image sensor used for testing

12.1.2 Frontside-Thinned Front-Illuminated CCDs

Because the optical absorption in SiO_2 becomes noticeable in the DUV,[1] ionization damages and UV-induced effects in the SiO_2 layer of the CCD are relevant concerns when using CCD sensors for DUV imaging. In terms of the ionization damages, the two major effects are the oxide-trapped charge and interface trap generation.[2] In addition, many UV degradation artifacts in Si-SiO_2 devices are linked to the UV-induced effects in the SiO_2 layer and at the interface. It is possible to improve the CCD's radiation hardness by optimizing the oxide thickness and processing conditions. Based on these reasons, it appears that thinning the oxide on top of the photosensitive regions of a CCD sensor can mitigate the occurrence of UV degradation artifacts. A reduced oxide volume lessens the interaction of radiation with the oxide, thus lowering the probability of UV-induced damages and decreasing the net absorption in SiO_2. If the oxide is sufficiently thin, there is be no ionization damage to speak of [14].

Two types of front-illuminated linescan CCD sensors with thinned oxide over the photodiode-based pixels are fabricated and tested in this investigation. The sensors employ a two-phase buried-channel CCD shift register and feature 512 14-μm square pixels. The photosensitive region (i.e., the pixel) is a p-n junction photodiode with the oxide overlayers partially removed to improve the UV sensitivity. Figure 12.2 is a simplified architectural view of the photodiode-based linear CCD image sensor.[3]

The distinction between the two types of CCDs is the different overlying oxide thicknesses and Si-SiO_2 interface qualities in the photosensitive regions. These sensors have oxide thicknesses of approximately 1398 Å (Sample-A) and 724 Å (Sample-B).[4] The device with the thinner oxide, Sample-B, is

---

[1] The optical properties of SiO_2 were described in Chap. 5.
[2] Ioniziation damages in SiO_2 were described in Chap. 7.
[3] The CCD samples are based on the modified IL-P3 CCD sensor, supplied by DALSA Corporation.
[4] The oxide thicknesses are measured by ellipsometry on the photodiode test structures that are fabricated following identical procedures as those for the

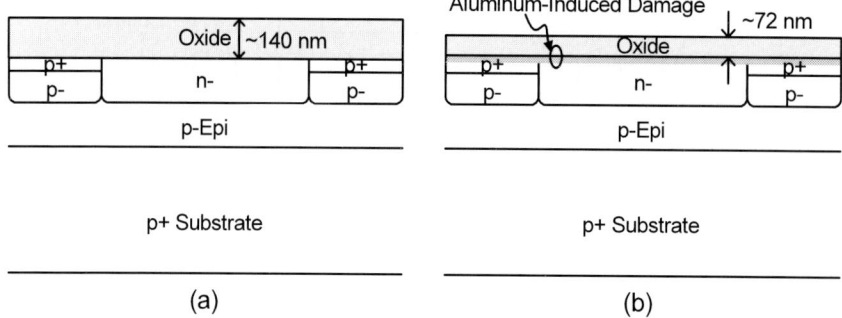

Fig. 12.3. A cross-section of the etched photosite for the thinned frontside-illuminated linescan CCDs under investigation. (**a**) Sample-A has thicker oxide and better interface quality. (**b**) Sample-B has thinner oxide but a poorer interface quality due to possible aluminum spiking. The dimensions are not drawn to scale

fabricated by using a different process recipe that exposes the Si surface to aluminum, and possibly other impurities, during the fabrication. The Sample-A sensor with the thicker oxide is not similarly exposed. Thus, the Sample-B sensor exhibits a poorer Si-SiO$_2$ interface quality. A cross-section of the two types of pixels is illustrated in Fig. 12.3.

Both types of thinned CCD samples have a higher PRNU at visible wavelengths than that of a regular non-thinned CCD sensor. The higher PRNU is likely to precipitate from the physical non-uniformities that are introduced by the oxide etching process. Sample-B has an even higher PRNU than that of Sample-A due to an additional etching step, which resulted in greater non-uniformity at the interface and on the surface of the photodiode region. The measurements of PRNU in the visible region for the thinned CCDs indicate that the PRNU is a function of the wavelength. This is perhaps due to the local differences that are introduced in the pixels during the device processing, so that each pixel contains varying quantities and types of defects or precursors. Thus, each pixel exhibits a slightly different responsivity at different wavelengths to result in a wavelength-dependent PRNU. These dispersions in the pixel property render the individual pixels to varying degrees of radiation damage.

With the exception of the photodiode and the output node regions, all the other components (e.g., the CCD registers) of this thinned front-illuminated CCD design are shielded from radiation by a top metal layer. Therefore, it is adequate to focus on the modifications occurring in the photodiodes

---

corresponding thinned CCD sensors. Photodiodes are used because of the difficulty of making direct measurements of the oxide thickness of CCD sensors. To provide a means of comparison, the overlying oxide thickness at the pixel of a conventional, unetched front-illuminated linear CCD is 2–3 µm.

(or pixels) and the output node of the sensor, when the radiation-induced damages are examined.

12.1.3 Laser Exposure Conditions

The thinned CCD sensors are exposed to a 157 nm F_2 excimer laser at intensities ranging from 0.2 to 10 pJ/pulse (these values correspond to the laser fluence hitting the CCD sensor). The pulse frequency is 100 Hz and the pulse width is 15 ns. Extrinsic QE measurements are first conducted at a low laser intensity of 0.2 pJ/pulse to ensure that the sensors are not in saturation, and the response is monitored for approximately one hour. Next, the sensors are exposed to a higher laser intensity of 10 pJ/pulse for the accelerated degradation testing. In the experiment, the CCD output signal, V_{OS}, is measured. Table 12.1 outlines the sequence of measurements for this experiment, whose results will be discussed in the next sections.

Table 12.1. The sequence of measurements for the 157 nm experiment

	Device Type	Device ID	Experiment Description
Part A	Sample-A	A1	(1) 157 nm response measurement at laser intensities of 1.6 pJ/pulse and 0.2 pJ/pulse at 100 Hz. (2) Monitor response over a 125 min. period at 0.2 pJ/pulse at 100 Hz
Part B	Sample-B	B1	157 nm response measurement over a 30 min. period at 0.2 pJ/pulse at 100 Hz
Part C	Sample-A	A2	High intensity exposure of un-powered sensor for 1 hour at 10 pJ/pulse at 100 Hz
Part D	Sample-B	B2	High intensity exposure of un-powered sensor for 15 min at 10 pJ/pulse at 100 Hz

The measured output signal, V_{OS}, is primarily comprised of two elements: the response signal, V_{signal}, and the dark signal, V_{dark}, where

$$V_{OS} = V_{signal} + V_{dark} . \tag{12.1}$$

To determine the extrinsic QE, the measured V_{dark} is subtracted from the measured V_{OS}. This difference, which is denoted by V_{signal}, is converted to an equivalent value for the number of electrons that are generated by the incident radiation. Then, the number of electrons is divided by the number of incident photons (calculated from the laser intensity measured at the CCD sensor) to obtain the extrinsic QE.[5] The extrinsic QE data, given in Sect. 12.2, correspond to the 157 nm radiation only; the response of the CCD to miscellaneous

---
[5] The relationship between the extrinsic QE and the various parameters were discussed in Chap. 2.

sources of radiation (e.g., the red component of the F_2 laser output, and the 157 nm induced fluorescence of the fused silica plates) has been deducted in the extrinsic QE calculations. For simplicity, in the following discussion, all citations and references of QE will imply extrinsic QE.

12.2 Experimental Results

12.2.1 Response Measurement of Sample-A at 157 nm

To evaluate the responsivity and extrinsic QE of a thinned front-illuminated linear CCD sensor to 157 nm radiation, a Sample-A sensor (identified as device A1) is irradiated with a F_2 excimer laser at intensities of 0.2 pJ/pulse to 1.6 pJ/pulse and a repetition rate of 100 Hz. The sensor receives a total of 751 200 laser pulses. Sufficient attenuation is used to ensure the CCD does not saturate during the response measurement. Brief interruptions of the irradiation on the CCD are required to adjust the attenuation and to measure the laser output energy. The results of the extrinsic QE measurement at 157 nm for Sample-A are represented in Fig. 12.4. Clearly, the thinned front-illuminated CCD sensor is sensitive at 157 nm; it offers a considerable improvement compared to the conventional front-illuminated CCDs which are not responsive to 157 nm radiation at all.

The measurements indicate that the CCD response, V_{OS}, depends on the laser frequency and laser intensity. This is logical because a higher laser intensity is equivalent to a larger number of photons, hitting the CCD per

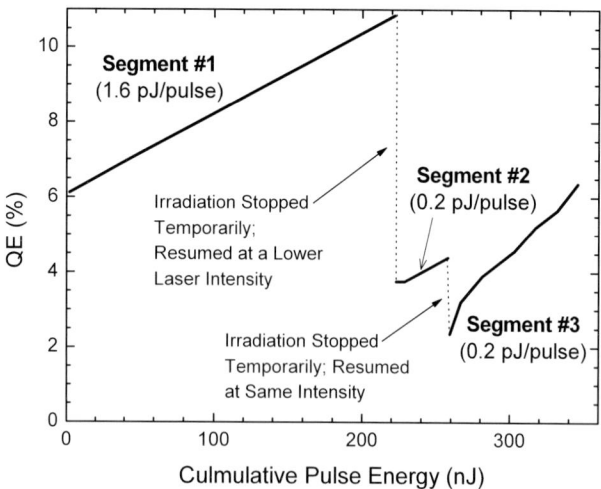

Fig. 12.4. The temporal fluctuations of extrinsic QE at 157 nm for Sample-A, measured at F_2 laser intensities of 1.6 pJ/pulse and 0.2 pJ/pulse

laser pulse.[6] Subsequently, more e-h pairs are generated in the Si layer and the photoresponse increases with an increase in the laser intensity. Similarly, a higher laser frequency means more photons are incident on the CCD per unit time; thus, the CCD photoresponse increases with the frequency.

A more striking observation from the data in Fig. 12.4 is the fluctuations in the extrinsic QE as a function of the cumulative UV dose. The QE at 157 nm increases with the continued exposure, but the QE decreases after the laser exposure is momentarily interrupted. The QE enhancement is evident in Segment #1 of Fig. 12.4. The CCD output signal increases from a small value to the saturation level after a continuous irradiation at 1.6 pJ/pulse for half an hour, indicating that the extrinsic QE increases over time. A similar enhancement of the extrinsic QE with the continuous exposure at a lower F_2 laser intensity is illustrated in Segment #2 and Segment #3 of Fig. 12.4.

Although the radiation-induced dark current generation is a potential contributor to the observed rise in the CCD signal, such a large increase in the CCD signal, up to the saturation level, is unlikely to be the result of the dark current generation alone. This argument is supported by the changes in the dark current level, measured in this part of the experiment. Prior to the 157 nm irradiation, the sensor has V_{dark} of 8 mV at 100 Hz ($J_{\text{dark}} \simeq 2\,\text{nA/cm}^2$). After the 0.2 pJ/pulse exposure period, V_{dark} of the exposed pixels increases to an average of 32 mV ($J_{\text{dark}} \simeq 8\,\text{nA/cm}^2$). Although the dark current is modified by the radiation, but this radiation-induced dark current generation is not large enough to saturate the CCD sensor. Consequently, it is logical to conclude that the CCD's responsivity is indeed modified by the 157 nm irradiation.

Figure 12.4 also shows that the extrinsic QE decreases whenever the irradiation is interrupted. For example, at the end of Segment #2, the QE drops by about 2%, after the irradiation has been terminated for 5 min. A similar decrease in the QE is observed after a two hour termination period, despite the measurements are conducted using the same laser intensity (0.2 pJ/pulse); in this case, ΔQE $\simeq -2\%$ after 2.25 hours of idle time.

It is suspected that the QE fluctuations are related to the ionization damages in the SiO_2 layer. In Sect. 7.3, it was stated that at low dose levels, the radiation-induced flatband voltage shift, ΔV_{FB}, in MOS structures is predominantly due to the oxide-trapped charges. It is reasonable to speculate that a similar situation holds for the CCD, where oxide charges are generated in the CCD pixels by a low intensity irradiation at 157 nm. The accumulation of sufficient oxide or interface charges affect the electrical properties of the p-n junction photodiode region, which subsequently affects the CCE and the charge capacity of the pixel. As a result, the extrinsic QE changes with the exposure dose. Other potential processes, intervening with the QE stability

---

[6]The laser intensity, in units of *Joule per pulse*, can be converted to units of *photon per pulse* after normalizing by the photon energy. Thus, a higher laser intensity implies a larger quantity of photons in each laser pulse.

at 157 nm, include UV-induced absorption changes in the SiO_2 due to the formation and restoration of color centers and interfacial modifications. A detailed analysis of the possible mechanisms that are responsible for the QE fluctuations is presented in Sect. 12.3.1.

12.2.2 Response Measurement of Sample-B at 157 nm

The 157 nm extrinsic QE of a Sample-B CCD sensor (identified as device B1) is measured by exposing the sensor to a F_2 excimer laser at an intensity of 0.2 pJ/pulse and a repetition rate of 100 Hz. The CCD's response to the 157 nm irradiation is monitored for approximately half an hour and the sensor receives a total of 264 000 laser pulses. As stated in Sect. 12.1.2, the Sample-B sensor has a high PRNU and an intrinsically high dark current, possibly due to Sample-B's poor Si-SiO_2 interface quality. Prior to the 157 nm irradiation, the average V_{dark} of the pixels in the B1 sensor is approximately 230 mV at 100 Hz ($J_{dark} \simeq 57.13\,nA/cm^2$).

The data from the extrinsic QE measurement of Sample-B at 157 nm are plotted in Fig. 12.5. It can be seen that this type of CCD sensor offers a higher extrinsic QE at 157 nm because of the decreased UV absorption in the thinner overlying oxide layer. This is evident by comparing the QE of Sample-B in Fig. 12.5 to the QE of Sample-A in Segment #2 and Segment #3 of Fig. 12.4. Similar to the QE of Sample-A, the QE of Sample-B increases as a function of the cumulative 157 nm dose with the continuous irradiation at 0.2 pJ/pulse. The QE increases at a constant rate throughout the exposure

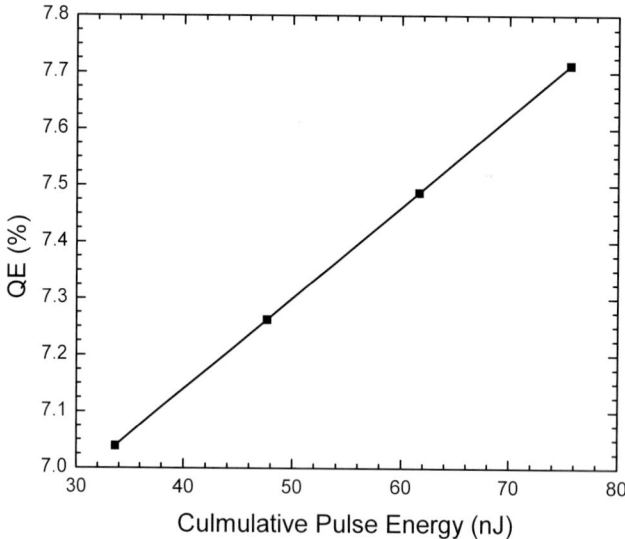

Fig. 12.5. The 157 nm extrinsic QE of Sample-B measured at a F_2 laser intensity of 0.2 pJ/pulse

period. The origins of the UV-induced QE enhancements are discussed in Sect. 12.3.1.

During the response measurement of Sample-B, changes in the dark current are observed. Peculiarly, the dark current decreases after the sensor is briefly irradiated at a F_2 laser intensity of 1.4 pJ/pulse; the situation is described here. To measure the 157 nm response of Sample-B, the F_2 laser irradiation begins at an intensity of 0.2 pJ/pulse. However, the sensor is initially unresponsive to the 157 nm irradiation, as the measured V_{OS} is the same as the intrinsic V_{dark} ($\simeq$230 mV); thus, the QE is almost 0%. To acquire a noticeable CCD response, the laser intensity is increased to 1.4 pJ/pulse; at this intensity, the sensor output saturates immediately. Surprisingly, when the laser intensity is lowered to the previous setting of 0.2 pJ/pulse, the sensor becomes responsive with a QE of approximately 7%. Furthermore, after the 2 min. irradiation at 1.4 pJ/pulse, V_{dark} reduces to almost 0 mV (i.e., the amplitude of V_{dark} is hardly detectable on the oscilloscope). Subsequently, the F_2 laser intensity is maintained at 0.2 pJ/pulse, and response measurement is performed. Following the 0.2 pJ/pulse irradiation segment and a 2.75-hour idle period, V_{dark} of the exposed pixels averages around 92 mV. These changes indicate that the dark current is altered by the low intensity F_2 laser irradiation. The variation in the dark current of Sample-B due to the low intensity 157 nm exposure is summarized in Table 12.2. An explanation for the changes in the dark current is attempted in Sect. 12.3.2.

Table 12.2. The changes in the dark current of the Sample-B sensor after a series of low intensity 157 nm F_2 laser exposure

	Sequence of Measurements	V_{dark} at 100 Hz	J_{dark} at 100 Hz
1.	Before 157 nm irradiation	230 mV	57.13 nA/cm^2
2.	After 157 nm irradiation at 1.4 pJ/pulse for 2 min	$\sim$0 mV	$\sim$0 nA/cm^2
3.	After 157 nm irradiation at 0.2 pJ/pulse for 30 min., and idle for 2.75 hr	92 mV	22.85 nA/cm^2

The initial changes in the B1 sensor's response to the 1.4 pJ/pulse irradiation can be interpreted as follows. First, Sample-B has an intrinsically large dark current, which usually suggests a large density of interface defects. When Sample-B is initially irradiated at 0.2 pJ/pulse, the photogenerated electrons in the Si layer are easily trapped at the interface states and cannot contribute to the QE. Subsequently, when the laser intensity is increased to 1.4 pJ/pulse, more carriers are generated in the Si layer. Perhaps, after the interface traps are filled, excess carriers are available to contribute to the carrier collection at the pixel. Thus, the CCD response increases and becomes saturated with the 1.4 pJ/pulse irradiation. Alternatively, the saturation of the CCD signal and the decrease in the dark current, after the 1.4 pJ/pulse irradiation, are linked to the UV-induced oxide charge generation (as discussed in Sect. 10.5). If the F_2 laser irradiation at 1.4 pJ/pulse generates a sufficient number of negative oxide charges, holes in the Si layer are attracted to the interface to passivate the interface traps. As a result, less interface trapping leads to a smaller dark current and a higher CCD response.[7]

12.2.3 Higher Intensity 157 nm Exposure on Sample-A: Accelerated Degradation Testing

The sensors are exposed to a higher F_2 excimer laser intensity of 10 pJ/pulse for accelerated degradation testing. The pixels in a Sample-A sensor (identified as device A2) receive 486 000 laser pulses at 10 pJ/pulse for a total exposure time of 70 min. At this high laser intensity, the CCD sensor is in saturation. Typically, high intensity irradiation causes latch-up in semiconductor devices, since a substantial amount of charge is injected into the device substrate. To prevent severe damage to the CCD camera and the measurement equipment, the camera is powered off during the high laser intensity irradiation.

The following is the sequence of exposure conditions. First, the unpowered CCD is irradiated at 10 pJ/pulse for a period of time. Then, the laser intensity is briefly lowered to 0.2 pJ/pulse, and V_{OS} is recorded. After the measurement, the laser intensity is returned to 10 pJ/pulse to continue the high intensity exposure experiment. This sequence is repeated twice. The change in V_{OS} (at 0.2 pJ/pulse, 100 Hz) as a function of the cumulative exposure at 10 pJ/pulse is monitored to evaluate the impact of a high intensity F_2 laser irradiation on the CCD. The experimental results for Sample-A are summarized in Fig. 12.6. V_{OS} increases with the cumulative exposure at 10 pJ/pulse, and the sensor becomes saturated at the end of the exposure. More specifically, for a pixel in the A2 sensor, V_{OS} (at 0.2 pJ/pulse) changes from 22 mV (prior to the high intensity exposure) to a saturation amplitude of 1.076 V (after a 10 pJ/pulse irradiation for 70 min.). There is at least a 48 times amplification in the output signal, indicating that the sensor's properties are modified by the high intensity radiation. For this particular device, 1.076 V is the saturation voltage; thus, the actual increase in V_{OS} is potentially larger than the measured saturation value.

As stated in (12.1), V_{OS} is the sum of V_{signal} and V_{dark}. The significant increase in V_{OS} (at 0.2 pJ/pulse) after the 10 pJ/pulse exposure can be attributed to an increase in V_{signal}, or V_{dark}, or both. Based on the results presented in Sect. 12.2.1, an increase in V_{signal}, associated with an increase in the QE, is possible with DUV exposure. However, a DUV-induced enhancement in QE is only very moderate in magnitude; thus, it is doubtful that

---

[7] This oxide charging effect is schematically recorded in Fig. 12.9, and will be discussed in greater detail in Sect. 12.3.1.

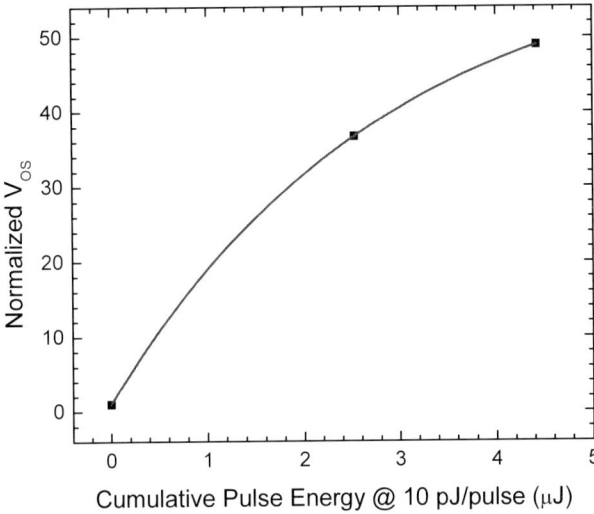

Fig. 12.6. The normalized CCD output signal, V_{OS}, of Sample-A (measured intermittently at 0.2 pJ/pulse) as a function of the culmulative exposure at a higher F_2 laser intensity exposure of 10 pJ/pulse. The values are normalized to the level prior to 10 pJ/pulse exposure. The *solid* line displays an exponential fit to the data in the form, $y = A_1 * \exp(x/t_1) + y_0$

the $\geqslant 48$ times increment in V_{OS} is solely caused by the changes in V_{signal}. If V_{signal} is assumed to increase by 48 times, then equivalently, the QE increases from 4% (before the high intensity exposure) to 192%; such a drastic upsurge in QE is highly disputable. Furthermore, Sect. 12.2.1 showed that the enhancement in QE is not permanent, and is accompanied by a QE reduction when the irradiation is interrupted. However, a reduction in V_{OS} is not detected in this part of the experiment after a momentary interruption, further supporting the speculation that the observed increase in V_{OS} is not strongly correlated to V_{signal}.

Given the fact that the dark current generation is a common consequence of radiation damage [14],[8] it is highly probable that the increment in V_{OS} with the higher intensity 157 nm irradiation is derived from the DUV-induced increase in the V_{dark}. This idea is further substantiated by the observation that the dark current of Sample-B increases drastically after the 10 pJ/pulse irradiation (see the next section). Additionally, the discussion in Sect. 7.3 suggests that high radiation doses often involve the creation of interface states. It follows, then, that the higher laser intensity irradiation can degrade the Si-SiO$_2$ interface quality to lead more carrier trapping at the interface states, and thus a higher dark current. More thorough measurements will be required in future investigations to confirm these analyses.

---

[8] See Chap. 8 for a discussion of radiation damage of CCDs.

12.2.4 Higher Intensity 157 nm Exposure on Sample-B: Dark Current Measurement

An accelerated degradation test is conducted on a Sample-B CCD sensor (identified as device B2) by irradiating the sensor at a higher F_2 laser intensity of 10 pJ/pulse for a total of 90 000 laser pulses. To avoid latch-up and severe damage to the device, the CCD camera is powered off during the high intensity exposure period. The dark current of Sample-B is measured intermittently throughout the exposure period to provide a means for examining the damages induced by the high intensity DUV irradiation.

The key observation is that the dark current density of Sample-B increases almost exponentially as a function of the cumulative 157 nm dose, as revealed in Fig. 12.7. In addition, Sample-B exhibits a larger increase in the DUV-induced dark current after the irradiation at 10 pJ/pulse, than Sample-A. For instance, a pixel in the Sample-B sensor undergoes a 58 times increase in the dark current after only 15 min. of irradiation at 10 pJ/pulse. Therefore, this type of sensor, with a thinner overlying oxide and an inferior interface quality, is more vulnerable to the DUV-induced dark current instability. Because a poorer interface is characterized by a higher intrinsic defect concentration at the $Si-SiO_2$ interface, these defects can accelerate the radiation-induced reactions that are responsible for the interfacial damage and the dark current generation.

Another observation is that the high intensity F_2 laser irradiation aggravates the DSNU. Because the fabrication process can introduce varying

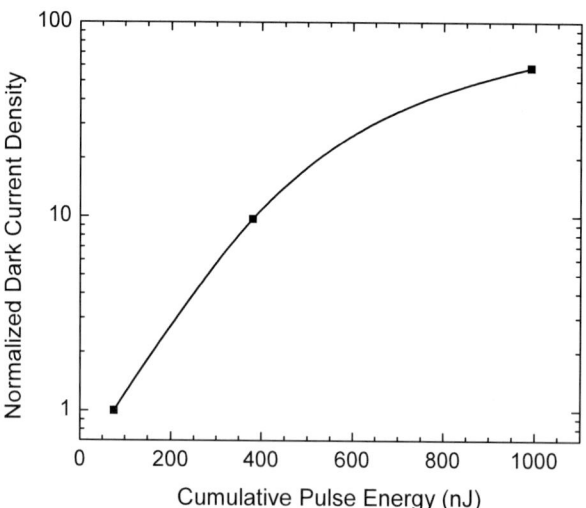

Fig. 12.7. The normalized change in the dark current density as a function of the 157 nm exposure at 10 pJ/pulse for a pixel in Sample-B. The values are normalized to the level, prior to the exposure

quantities and types of defects in each pixel, it is expected that the DUV radiation has an inharmonious effect on each pixel. In addition, the radiation damage sites typically occur as isolated areas in the material and are not uniformly distributed [18]. As a result, the DUV-induced damage leads to a wider dispersion of the dark current across the CCD sensor, and thus a larger DSNU.[9]

It is desirable to investigate the impact of the higher intensity DUV exposure on CCD's responsivity and QE. An attempt is made to measure the 157 nm extrinsic QE of Sample-B, following the 10 pJ/pulse exposure period. However, V_{OS} of the exposed pixels is saturated even when the measurement is taken at a low laser intensity of 0.2 pJ/pulse at 100 Hz. Thus, the QE cannot be accurately computed. It is probable that the drastic increase in dark current after the 10 pJ/pulse exposure overwhelms the V_{OS} to result in a saturated response; this observation corroborates the explanation given in Sect. 12.2.3 for the Sample-A sensor.

Because the 157 nm photons are absorbed very close to the interface in the Si layer and because a dependence of the dark current on the interface quality is observed, it is reasonable to postulate that the changes in the dark current are a result of the DUV-induced modifications in the Si-SiO$_2$ interface. The mechanisms behind the dark current instability will be explored in Sect. 12.3.2.

12.3 Analysis of DUV Damage Mechanisms

Clearly, the experimental outcomes demonstrate that the oxide thickness and interfacial quality have a significant influence on the DUV responsivity and stability of a CCD sensor. A CCD with a thinner overlying oxide yields a higher extrinsic QE, and those with an inferior interface exhibit a larger dark current instability with the DUV irradiation. Although visible physical damage is not detected on the sensor surface after it is exposed to the F$_2$ excimer laser, the QE and dark current data suggest that the DUV-laser-induced damages occur nonetheless. These damages are likely to give rise to defect states within the band-gap of SiO$_2$ and cause structural modifications of the interface, leading to the formation of traps and leakages. Moreover, the process of ionization damage, discussed in Chap. 7 and Chap. 8, can also contribute to the DUV-induced changes in the CCDs. Radiation damage on the peripheral circuits is not expected, since all the other sections (except the photodiode and output node) of the CCD sensor are covered with a metal shielding layer, inhibiting direct radiation damage in these regions. Justifications and explanations for the CCD behavior at 157 nm are presented, in light of the theories and findings on the material-and-radiation interactions that were presented in the preceding chapters.

---

[9] Refer to Sect. 2.3.3 for information on DSNU.

12.3.1 Analysis of 157 nm Response Measurements

If the response measurement of the two thinned CCD sensors with different overlying oxide thicknesses are compared, the thinner oxide sensor offers a higher responsivity in DUV. This agrees with the theoretical expectation that the thinner oxide absorbs less DUV radiation for any given irradiation time than the thicker oxide device. The result is that more photons arrive in the Si layer to contribute to the QE at 157 nm. In addition, studies have shown that the radiation-induced oxide charges vary superlinearly with the oxide thicknesses [49].[10] These reasons provide a justification for the observed dependence of the CCD's DUV response on the oxide thickness.

When the CCD samples are exposed to 157 nm radiation at an intensity of 0.2 pJ/pulse, the extrinsic QE fluctuates with the cumulative exposure dose. The extrinsic QE is enhanced by the continuous exposure, but decreases when the irradiation is interrupted momentarily. Ionization damage in the SiO_2 layer is conceivably one of the mechanisms responsible for the extrinsic QE fluctuation. At low exposure dose, the generation of oxide-trapped charges is the dominant consequence of ionization damage [49].[11] Thus, low laser intensity irradiation of the CCD at 0.2 pJ/pulse can cause accumulation of sufficient oxide or interface charges to affect the electrical properties of a p-n junction photodiode; this can subsequently affect the charge capacity of the pixel and introduce potential traps in the photosensitive region. As a result, the extrinsic QE changes with DUV exposure.

DUV laser induced effects are also a contributing factor to the QE instability at 157 nm. These processes include the DUV-induced absorption changes in SiO_2 due to the formation and restoration of the color centers, DUV-induced oxide charge generation, DUV-induced bond or structural rearrangements, and interfacial modifications. According to the literature that was reviewed in Chap. 10, some of these degradation artifacts are time- or dose-dependent, which can trigger fluctuations in the CCD characteristics with exposure dose.

The UV-induced oxide charging effect provides additional insight on the QE fluctuations. Figure 12.4 demonstrates that the DUV-induced increase in the QE is partly reversible, if the irradiation on the CCD sensor ceases. This behavior points to a temporary charging effect. As discussed in Sect. 10.5, DUV-induced oxide charging is a dose-dependent phenomenon, where the fixed oxide charge density, N_f, in SiO_2 varies as a function of the cumulative DUV dose (see Fig. 10.11 and Fig. 10.12). Moreover, the manner in which N_f changes with the DUV exposure is dependent on the quality and type of the SiO_2 film [63]. According to basic semiconductor theory, the charge accumulation in SiO_2 attracts carriers in the Si layer to the Si-SiO_2 interface; as a result, the electrical properties and trapping dynamics at the interface

---

[10] The dependence of the radiation-induced oxide charges on the oxide thickness was described in Sect. 7.2.

[11] See Chap. 7 for a discussion of ionization damage.

are altered. Depending on the nature of the charges in the SiO_2 layer, these UV-induced oxide charges has either a positive or negative contribution to the QE at 157 nm.

The variations in the induced absorption of SiO_2 and the charge accumulation in the SiO_2 with the DUV dose, along with other photoinduced effects, have a significant impact on the DUV stability of the CCD sensors. These DUV-induced effects alter the optical and electrical properties of the oxide and interface layers, resulting in both temporary and permanent shifts in the device performance. In the following sections, more specific processes are proposed for the enhancement and reduction of the extrinsic QE at 157 nm.

12.3.1.1 Reasons for QE Enhancement

The enhancement in extrinsic QE with continuous 157 nm exposure, shown in Fig. 12.4, can be attributed to the following mechanisms: the ionization reactions in the SiO_2, the multi-photon absorption processes, the DUV-induced absorption in the SiO_2, the DUV-induced oxide charge modification, and the structural modifications.[12]

Mechanism #1.1: Because the photon energy of the F_2 excimer laser is close to the optical band-gap energy of the a-SiO_2 and that TPA events are possible with the high peak power of the excimer laser, a portion of the incident 157 nm photons are absorbed by the SiO_2 layer and result in the photogeneration of e-h pairs by ionization. Electrons in the SiO_2 conduction band can drift to the Si layer, where they add to the pool of photogenerated carriers in the Si. As a result, carrier collection in the pixel increases, which translates to a higher extrinsic QE. This mechanism is also evident in the UV photodiodes that have been reported by Canfield et al. [67]. They have demonstrated that a percentage of the absorption events in the oxide produce carriers that contribute to the carrier collection in the Si layer, resulting in positive contribution to the QE of the photodiode [67].[13] Figure 12.8 provides a simplified schematic of the photogeneration of the e-h pairs in the SiO_2 and the subsequent contribution to collection of carriers in the Si layer, which are responsible for the extrinsic QE enhancement.

Mechanism #1.2: The increased carrier population in the mid-gap or intermediate states in the SiO_2, due to the multiple photon excitation processes, can cause QE fluctuations. This hypothesis can be reasoned as follows. First, it was established, in Chap. 5, that a-SiO_2 is characterized by the intermediate states in the band-gap due to the structural disorder in the network. Secondly, the optically-induced transition by the multiple photon absorption processes in SiO_2 is possible with the DUV excimer laser owing to the high peak power and coherency of the laser pulses [53].[14] These facts collectively

---

[12] These DUV-induced effects were presented in Chap. 9 to Chap. 11.

[13] The details of this investigation, by Canfield et al., will be presented in Chap. 13 (see Fig. 13.6).

[14] See Chap. 10.

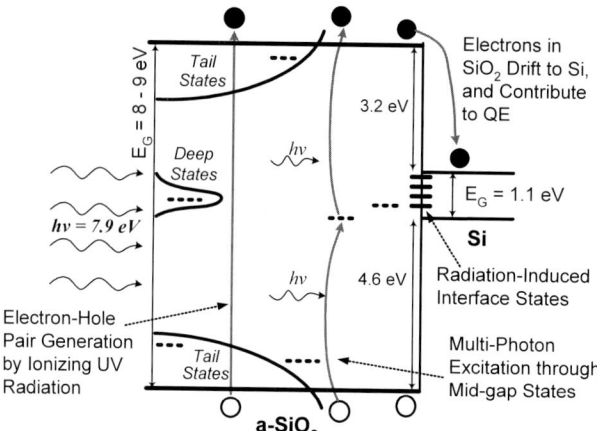

Fig. 12.8. A simplified illustration of the possible DUV-induced mechanisms responsible for the QE enhancement in the CCD sensor with a F_2 laser irradiation. The e-h pair generation process in SiO_2 by the ionizing UV radiation, and the multi-photon absorption process in SiO_2 are depicted here. The numerical values for the energy band diagram are obtained from [51] (The band bending of the Si-SiO_2 structure is not shown here.)

suggest that the successive photon absorption, from the continuous DUV laser exposure, enables the electrons in the SiO_2 to be excited to increasingly higher energy states within the band-gap, and finally, into the conduction band. Since the carrier population in the intermediate states is increased by this process, there is a higher probability for the electrons to become excited into the SiO_2 conduction band, where they can contribute to the QE. Since the number of electrons that occupy the intermediate energy states is a function of the exposure history, the extrinsic QE is also a function of the exposure history. A pictorial representation of this mechanism is provided in Fig. 12.8.

Mechanism #1.3: The DUV laser induced formation and restoration of color centers in SiO_2 can cause fluctuation in the SiO_2 absorption intensity. This was seen in Sect. 10.3.4, where that the UV-induced absorption intensity of SiO_2 at 6.4 eV (193 nm) increases with the continuous ArF (193 nm) excimer laser irradiation, but decreases when the laser is turned off [57]. This behavior, exhibited in Fig. 10.5, resembles the fluctuation pattern in the 157 nm extrinsic QE in Fig. 12.4. If the F_2 excimer laser induces similar variations in the absorption intensity of SiO_2 at 7.9 eV (157 nm), then a temporal increase in the number of photons that are absorbed in the oxide with continuous exposure can temporarily increase the QE of the CCD. This is due to the fact that an increasing number of electrons are generated in the SiO_2 layer, which then drift to the Si layer to add to the carrier collection in the pixel (as described in *Mechanism #1.1*). In this manner, the increasing

SiO$_2$ absorption, due to the color center formation, brings about an extrinsic QE enhancement with the continuous DUV exposure.

Mechanism #1.4: The F$_2$ laser induced decrease in the SiO$_2$ absorption at 157 nm and the bleaching of the VUV absorption edge, reported in Sect. 10.4.3, offer a different interpretation for the QE enhancement in CCDs. As demonstrated in Fig. 10.8 and Fig. 10.10, there is a decrease in the optical absorption of SiO$_2$ at 157 nm and at the absorption bands above 7.5 eV (near the VUV absorption edge) after a F$_2$ laser irradiation [62]. A reduced oxide absorption allows more 157 nm photons to penetrate into the Si layer, where they generate e-h pairs that directly contribute to the carrier collection of the pixel. Consequently, the extrinsic QE increases with the DUV exposure.

Mechanism #1.5: DUV-induced oxide charging presents another source of QE instability. Figure 12.4 is a reminder that the DUV-induced increase in the QE is partly reversible, when the irradiation on the CCD sensor is ceased; such reversible changes are typically linked to a temporary charging effect. In Sect. 10.5, it was established that the fixed oxide charge density in SiO$_2$ changes as a function of the accumulative DUV dose [63]. In addition, Sect. 5.2.3 showed that the fixed oxide charges were concentrated close to the Si-SiO$_2$ interface [41]. If the F$_2$ laser generates a sufficient density of negatively charged centers in the SiO$_2$ layer near the interface, the holes in the Si substrate are then attracted to the Si-SiO$_2$ interface to form an inversion layer, as illustrated in Fig. 12.9. This layer of holes can block and/or passivate the interface traps.[15] With fewer interface traps, fewer photogenerated electrons recombine and the extrinsic QE increases. The fluctuation in QE stems from the fact that the DUV-induced oxide charge density varies with the cumulative exposure dose. However, this interface trap passivation process holds only if enough negative oxide charges are generated. If there is an insufficient number of negative charges, they will merely deplete the Si layer underneath the interface to alter the channel potential, but the QE is unaffected. In addition to the DUV-induced oxide charge centers, the negative oxide charges can originate from the DUV-induced photoemission of electrons in the Si into the SiO$_2$ layer,[16] where they are trapped to charge the oxide negatively, and other electrically-active defects in the oxide.

Mechanism #1.6: The 7.9 eV photons from the F$_2$ excimer laser have sufficient energy to break or weaken the bonds, alter the bond angles, bond lengths, and even the molecular composition in the SiO$_2$ and in the Si-SiO$_2$ interface [19]. One consequence is the densification of SiO$_2$, as discussed in Sect. 10.1.2 [55]. UV laser-induced densification in SiO$_2$ not only causes the material density to increase, but it also triggers additional modifications of

---

[15] The process of the hole accumulation at the interface is very similar to the CCD inversion technique typically used for eliminating or reducing the surface dark current, which was discussed in Sect. 2.3.

[16] The photoemission of carriers in a Si-SiO$_2$ structure is discussed Sect. 8.2.1 and Chap. 9.

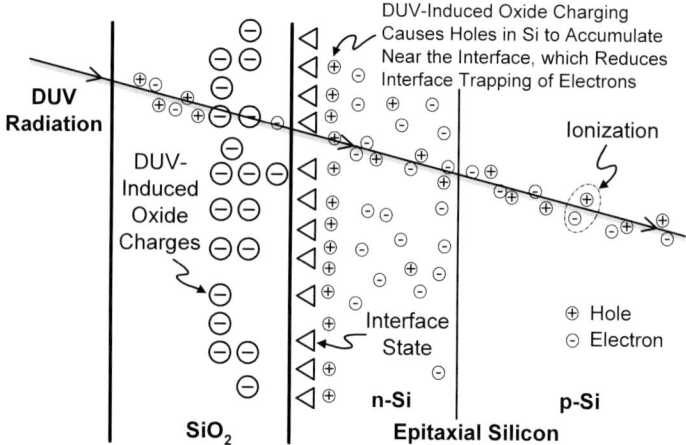

Fig. 12.9. A schematic illustration of the UV-induced oxide charging effect in the photodiode region to explain the extrinsic QE fluctuations at 157 nm for CCD. If UV photons induce enough negatively charged species in the oxide, holes accumulate at the interface and reduce the interface trapping of electrons; this results in an increase in the extrinsic QE and a decrease in the dark current

the material properties, including changes in the index of refraction (wavefront distortion), induced surface deformation of the SiO_2 layer, and stress-induced birefringence. Instances of DUV laser-induced structural changes in SiO_2 and in the Si-SiO_2 interface have been reported by Fiori and Devine [68] and by Lu et al. [65], respectively. However, these studies operated the DUV lasers at very high intensities, and the resulting structural modifications are mainly due to thermal effects. Since the QE measurement in this book's investigation involves low to moderate F_2 laser intensities, severe structural modifications are unlikely to occur. Instead, the F_2 laser induced rearrangement and breaking of the weak and/or strained bonds, in the SiO_2 or the Si-SiO_2 interface, are more probable. As the number of strained Si-O-Si bonds is decreased by the DUV-induced bond alteration, the DUV-VUV absorption, arising from these strained bonds (predominantly near 7.9 eV), decreases correspondingly [53]. The reduced oxide absorption results in an extrinsic QE enhancement. In addition, the DUV laser can modify the interface properties by triggering bond structure rearrangement in a manner that disfavors electron trapping by the interface states or promotes passivation of the interface traps. These events also stimulate an enhancement in the extrinsic QE at 157 nm.[17]

---

[17]The DUV-induced structural modifications were discussed in Chap. 10 and Chap. 11.

12.3.1.2 Reasons for QE Reduction after Interruption

The temporal QE reduction after a momentary interruption of the DUV laser exposure, depicted in Fig. 12.4, can be associated with the relaxation, recombination or restoration processes of the defect centers in the oxide and interface layers. Some of the potential mechanisms are described next.

Mechanism #2.1: In view of the multi-photon excitation process (described earlier in *Mechanism #1.2*), the electrons, that are already optically excited to the intermediate energy levels in the SiO_2 band-gap, can lose energy during the laser interruption and decay to a ground state. When the exposure is resumed, the electrons are at a ground state. Since a much smaller population of electrons is available at the mid-band levels, the excitation of the electrons into the conduction band by multi-photon absorption processes becomes very inefficient, immediately after the irradiation is resumed. Thus, a smaller extrinsic QE is observed.

Mechanism #2.2: The temporary charging effect, related to the DUV-induced oxide charge modification (proposed in *Mechanism #1.5*), can be a contributor to the QE reduction. If the DUV-induced oxide charges decay or relax into neutral species during the laser interruption, the electric field in the oxide weakens. This causes the holes in the inversion layer to drift away from the interface, and the passivation of the interface traps will not be present immediately after exposure resumes. As a result, the interface trapping of the photogenerated electrons occurs and causes a decrease in the extrinsic QE. An example of the decay in the UV-induced oxide charges was reported in Sect. 10.2.2, in which a fraction of the E' centers decays at room temperature after the DUV irradiation and is believed to be due to the recombination of the E' centers with the H_2 molecules in the SiO_2 [59]. This observation corroborates the mechanism that is proposed here.

Mechanism #2.3: The interface trapping dynamics can change, when the laser irradiation terminates, to contribute to the QE fluctuations. It is supposed that the interface traps are fully occupied during the continuous DUV radiation. However, the trapped carriers are not a permanent feature; they can slowly detrap [14]. Thus, the trapped charges can be released from the interface states during the interruption. When the exposure resumes, some of the photogenerated electrons are trapped by the newly available interface states, leading to a reduction in the extrinsic QE.[18]

Mechanism #2.4: If the changes in the DUV-induced absorption in SiO_2, illustrated in Fig. 10.5, holds for the case of the F_2 excimer laser irradiation, then the absorption intensity of SiO_2 at 7.9 eV (157 nm) decreases when the laser is turned off [57]. The reduced SiO_2 absorption implies that more photons strike the Si layer. However, since the penetration depth of the 157 nm photons in Si is very shallow[19] and since the interface traps are likely to

---

[18] The possible detrapping mechanisms were given in Sect. 5.2.3, Sect. 6.2.2 and Sect. 8.2.5.

[19] The optical properties of Si were presented in Chap. 4.

be unoccupied after the laser interruption (based on *Mechanism #2.3*), the electrons in the Si, which are photogenerated close to the Si-SiO$_2$ interface, will become trapped by the interface states. The resulting loss of carriers, due to the interface trapping, leads to a drop in the QE.

Due to the complexity of the mechanisms, it is difficult to conclude which is the dominant process responsible for the QE fluctuations. It is expected that many processes occur concurrently. More extensive research and investigations are required to gain a better insight into the processes, to estimate the magnitude of the contributions of the various mechanisms, to precisely identify the underlying causes of the QE instability, and to generate a unified model of the DUV-induced degradation of CCD sensors.

12.3.2 Analysis of Dark Current Measurements

The 157 nm experimental results indicate that the radiation-induced changes in the dark current depend strongly on the Si-SiO$_2$ interface properties and on the UV fluence. CCDs with an inferior interface quality (Sample-B) have a more severe DUV-induced dark current instability. Also, a more intense UV laser irradiation results in a larger increase in the CCD dark current as the exposure dose accumulates. This dark current dependency on the radiation intensity can be interpreted by using the theory presented in Sect. 7.3, which states that different damage mechanisms are activated at different radiation doses. At a low radiation dose, the oxide charge generation is the dominant mechanism responsible for the radiation-induced shifts in the MOS device performance, but at intermediate doses, the interface state generation becomes dominant [41]. These concepts are applied to explain the dark current behavior of CCD sensors after DUV irradiation.

12.3.2.1 Increase in Dark Current After Higher Intensity DUV Radiation

The substantial increase in the dark current density with the cumulative DUV dose (Fig. 12.7), when a CCD sensor is exposed to a higher F$_2$ laser intensity of 10 pJ/pulse, is logical from a qualitative standpoint. According to the basic CCD theory presented in Sect. 2.3, the dark current is a strong function of the interface state density, including unpassivated dangling bonds in the Si-SiO$_2$ interface [14]. Then, an increase in the dark current signifies the generation of more interface states, which is highly probable with the F$_2$ laser irradiation at 10 pJ/pulse for these reasons. First, it is recognized that interface state creation is the dominant ionization damage mechanism at intermediate doses of DUV radiation. Secondly, the DUV laser has sufficient energy to break and/or rearrange the bonds in the Si, SiO$_2$, and interface layers (Chap. 9 to Chap. 11). Therefore, the increase in dark current, observed after a higher F$_2$ laser intensity exposure, is essentially a consequence of the

DUV laser induced interface state formation. The exponential-like characteristic of the dark current density with increasing DUV dose is also justifiable, because many of the mechanisms and kinetics, governing the dark current, are associated with an exponential process.

12.3.2.2 Reduction in Dark Current After Lower Intensity DUV Radiation

The contrary behavior of dark current reduction, after a previously unexposed Sample-B sensor is briefly irradiated at a lower F_2 laser intensity of 1.4 pJ/pulse (Sect. 12.2.2), is attributed to various causes. One possibility is the passivation of interface states by holes, which are induced electrically by the creation of negatively charged centers in the oxide (as illustrated in Fig. 12.9). Another possibility is the DUV-induced structural modification at the interface, where the modified structure reduces the interface trapping, cutting back the dark current generation. Furthermore, the reduction in the dark current can be linked to a surface conditioning effect. It has been reported that many materials exhibit increased damage thresholds, following exposure to a one or two DUV laser pulses [18]. The initial pulse(s) at low repetition rates can be responsible for removing the surface contamination to provide a conditioning effect on the device surface. Thus, the device property is improved slightly and the dark current is reduced.

The causes of the DUV degradation behavior of CCD sensors are not limited to the damage mechanisms that are proposed in this chapter. A variety of other DUV laser induced mechanisms can also contribute to the DUV behavior of the CCD such as UV-induced oxidation, laser ablation, laser sputtering, laser annealing and amorphization, and laser-induced chemical etching. For instance, the DUV-induced oxidation modifies the oxide thickness, which ultimately affects the performance of CCD sensors. This DUV-induced oxidation occurs if an adequate concentration of oxidant is present in the ambient of the radiation chamber, and if the DUV fluence is sufficiently high. However, for this particular investigation, these effects are likely to be negligible because the laser fluence is too weak to activate these highly energetic mechanisms.

12.4 Post-157 nm Measurements

Measurements of the dark current density, visible extrinsic QE, and CCE after the 157 nm irradiation experiment indicate that these parameters are permanently altered by the DUV exposure. The dark current continues to change, declining even weeks after the last 157 nm exposure; the results are reported in Sect. 12.4.1 and Sect. 12.4.2. The extent of change in the post-DUV dark current appears to be contingent on the intensity of the DUV laser irradiation, as well as on the sensor's properties (oxide thickness and

178   12 CCD Measurements at 157 nm

interface quality). The visible QE and CCE are also modified by the 157 nm exposure, and these changes are explored in Sect. 12.4.3 to Sect. 12.4.5.

12.4.1 Dark Current Response of Sample-A After 157 nm Irradiation

In the 157 nm experiment, for the pixels in the A1 sensor that are exposed to a lower F_2 laser intensity of 0.2 pJ/pulse, they exhibited a smaller increase in the dark signal. For pixels in the A2 sensor that are exposed to a higher laser intensity at 10 pJ/pulse, a more substantial increase in the dark current density is observed. These changes in the dark current during the F_2 laser exposure were discussed in Sect. 12.2.

The first post-DUV dark current measurement of Sample-A is taken one week after the 157 nm exposure. For a pixel that is exposed to a F_2 laser at 0.2 pJ/pulse, identified as Pixel-A, the first post-DUV measurement indicates a five times enlargement in the dark current compared to its intrinsic (pre-DUV) value. For a pixel in A2 that is exposed to 10 pJ/pulse irradiation, identified as Pixel-B, the post-DUV dark current is about 42.5 times larger than its pre-DUV dark current value. The change in the dark current of Sample-A is monitored on a weekly basis over seven weeks, where the measurements are taken at room temperature and with the CCD camera running at an integration frequency of 100 Hz. The results are plotted in Fig. 12.10, where the post-DUV dark current density is normalized to the dark current of the respective pixel measured one week after the 157 nm exposure. Figure 12.11 provides an alternative view of the changes in the post-DUV dark current, where the dark current density for both pixels is normalized to the dark current of Pixel-B measured in week one. This figure shows that the DUV-induced dark current of Pixel-B is much larger than that of Pixel-A. Therefore, the impact of F_2 excimer laser irradiation on the post-DUV dark current is more dramatic after a higher intensity exposure.

Figure 12.10 and Fig. 12.11 reflect a gradual reduction in the dark current density after the 157 nm exposure. A comparison of the two pixels indicates that Pixel-B (previously exposed to a higher F_2 laser intensity) undergoes a faster and larger decay in its post-DUV dark current. The data set for Pixel-A and Pixel-B resemble almost a straight line in the log-log scale of Fig. 12.11. This implies that the changes in the post-DUV dark current in Sample-A are governed by some power-law processes of the form $y = ax^b$, where y is the normalized dark current density and x is the time (in weeks). A more in-depth investigation is necessary to determine the exact mechanisms that govern the behavior of the post-DUV dark current.

The decline in the post-DUV dark current suggests that some of the radiation damages or defects, responsible for the UV-induced dark current generation, have partially recovered after the irradiation, and is possibly related to a temporary oxide charging effect. Perhaps the UV-induced oxide charges slowly neutralize or recombine over time, altering the electrical properties

12.4 Post-157 nm Measurements   179

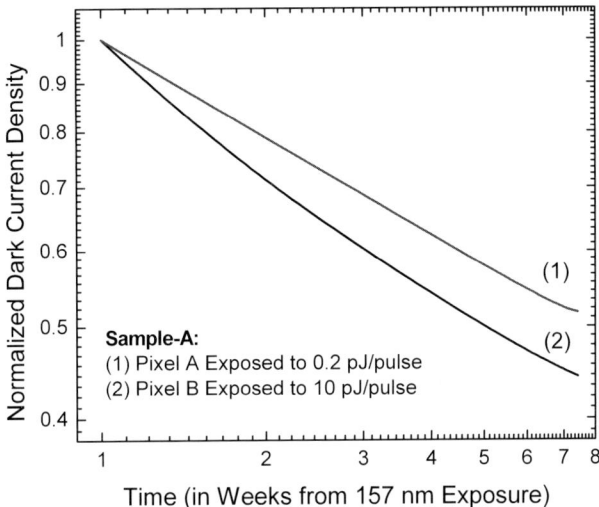

Fig. 12.10. The long term change in the dark current following a 157 nm exposure for two pixels in Sample-A that have been exposed to different F_2 laser intensities. The dark current density is normalized to the dark level of the respective pixel, recorded one week after the 157 nm exposure

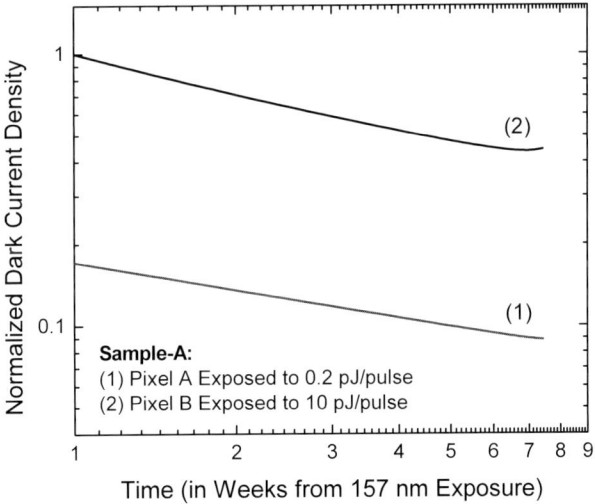

Fig. 12.11. The long term change in the dark current, following a 157 nm exposure, for two pixels in Sample-A that have been exposed to different F_2 laser intensities. The dark current density is normalized to the dark level of Pixel-B recorded one week after the 157 nm exposure

and the trapping dynamics of the interface. These reversible changes in the UV-induced dark current can also be due to the on-going relaxation or rearrangement of the radiation-induced interface traps, or the transformation of the DUV-induced defects in the SiO_2 layer. Another potential reason for the post-DUV dark current reduction is associated with the annealing process, where a trapped electron escapes from a shallow interface trap thermally at room temperature.[20] Nevertheless, since the extent of the dark current recovery is still far from the pre-DUV-irradiation level even after a period of two months, it is highly probable that radiation damage has occurred in the CCD to permanently modify the dark current characteristics. In summary, the UV-induced changes in the dark current can be partly attributed to a temporary oxide charging effect, and partly to a radiation damage effect that results in permanent changes in the CCD dark current.

12.4.2 Dark Current Response of Sample-B After 157 nm Irradiation

To study the post-DUV irradiation characteristics of the CCD sensors, the dark current of Sample-B is monitored on a weekly basis for eight weeks after the 157 nm experiment. Figure 12.12 represents the relative change in the post-DUV dark current of Sample-B for the two pixels that are previously exposed to different F_2 laser intensities, where Pixel-A corresponds to a pixel exposed to a lower F_2 laser intensity of 0.2 pJ/pulse and Pixel-B corresponds to a pixel exposed to a higher intensity of 10 pJ/pulse. The horizontal time axis is in units of weeks, where week 1 corresponds to the moment immediately after the termination of the 157 nm experiment. The vertical axis is the dark current density, normalized to the dark current level of Pixel-B measured at week 1. It can be seen that the dark current of Pixel-B is higher than the dark current of Pixel-A. Nevertheless, the post-DUV dark response for both pixels in Sample-B decreases gradually with time. The curves in the log-log plot of Fig. 12.12 are almost linear, implying a power relation, $y = cx^d$; here, y is the normalized dark current density and x is the time (in weeks). Thus, it appears that the changes in the post-DUV dark current in the eight weeks after the 157 nm exposure is governed by some power-law processes. The possible origins of the dark current reduction after the DUV exposure, described in Sect. 12.4.1, are also applicable here. However, the exact mechanism and processes are unknown at this stage.

If the post-DUV dark current behavior of the Sample-A and the Sample-B sensors are compared, a larger change in the post-DUV dark current density is observed for the Sample-B sensor (a thinner oxide and a poorer interface quality). This can be attributed to the inferior interface quality of Sample-B which is susceptible to more severe radiation-induced damages at the interface. In addition, the change in the post-DUV dark current with time displays

---

[20]This is analogous to the annealing of the CCD sensor, described in Sect. 8.2.5.

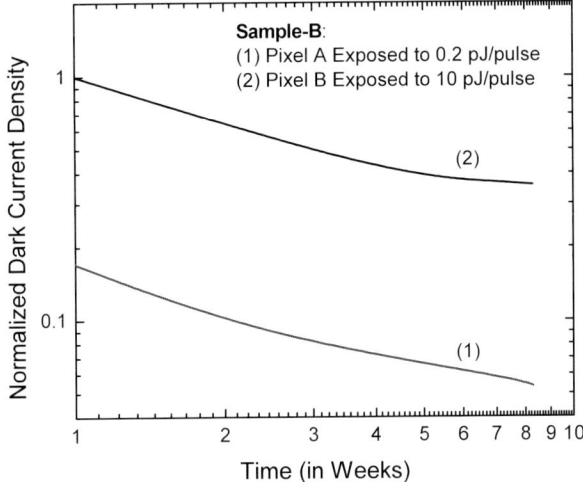

Fig. 12.12. The long term change in the dark current, following a 157 nm exposure, for two pixels in Sample-B that have been exposed to different F_2 laser intensities. The dark current density is normalized to the dark level of Pixel-B immediately after the exposure (i.e., at week 1)

slightly different trends between the two sensors. For example, there is a difference in the steepness (or slope) of the curves in Fig. 12.11 and Fig. 12.12. Possibly, these distinctions are related to the different oxide thicknesses and interface qualities of the two CCD sensors.

12.4.3 DUV-Induced Changes in Visible QE of Sample-A

The visible extrinsic QE of the thinned CCD sensors is measured before and after the 157 nm exposure, for visible wavelengths from 400 nm to 700 nm. The QE of Sample-A decreases across most of the visible spectrum after the 157 nm irradiation, with the largest QE reduction occurring at the shortest visible wavelength (400 nm). Similar visible QE behavior was observed for the pixels in both A1 and A2 sensors. These changes in the visible extrinsic QE are illustrated in Fig. 12.13 and will be examined next.

If the 157 nm irradiation induces an increase in the interface state density by breaking the weak or strained bonds at the interface, these UV-induced interface states probably remain even after the DUV exposure and can trap photogenerated carriers. Such trapping results in a degraded response and a poorer QE for the CCD sensor. Because the shorter wavelength photon has a smaller absorption depth in Si than a longer wavelength photon, the UV-induced interface traps are more likely to interact with the carriers that

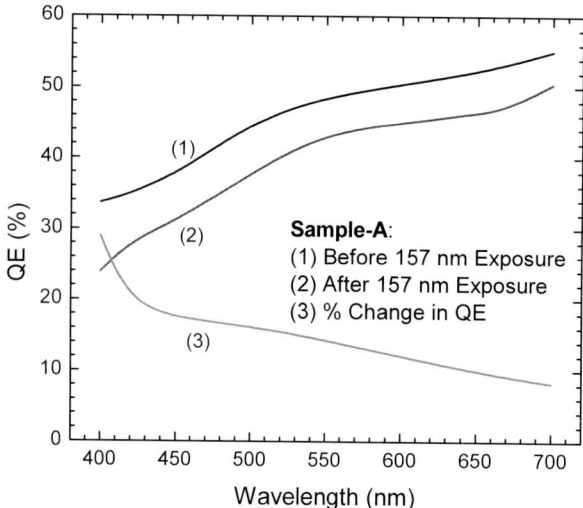

Fig. 12.13. The visible extrinsic QE of Sample-A before and after the 157 nm irradiation. The displayed QE values are determined by averaging the extrinsic QE for a number of sample pixels across the sensor. The measurements are performed at room temperature and at a CCD integration frequency of 15 kHz to 30 kHz

are generated very close to the interface.[21] Accordingly, these interface states have a larger impact on the extrinsic QE at the shorter visible wavelengths. This is reflected in Fig. 12.13 where a larger percentage of change in the QE is observed at the shorter wavelengths.

Alternatively, the decrease in the visible QE, following the 157 nm irradiation, can be viewed in terms of the UV-induced oxide absorption due to the defect or color center formations. It is known, from Chap. 5, that an assortment of defects can exist in a-SiO_2, and some of these defects or color centers are activated by the UV excitation to introduce optical absorption bands at visible wavelengths. An increase in the optical absorption of SiO_2 can have an impact on the visible extrinsic QE of the CCDs after the DUV irradiation.

12.4.4 DUV-Induced Changes in Visible QE of Sample-B

To evaluate the impact of the 157 nm irradiation on the visible response of the Sample-B CCD sensor with the thinner overlying oxide, the visible QE is measured before and after the 157 nm exposure experiment to provide a basis for the comparison and analysis. However, it is quite difficult to make accurate QE measurements for Sample-B because of the high degree of

---

[21] Refer to Chap. 4 and Chap. 7 for the optical properties and radiation effects of Si.

non-uniformity across the sensor. The larger DSNU and PRNU of Sample-B, compared to those of Sample-A, complicate the visible response measurements of Sample-B. Due to these non-uniformities, the pixels at the different regions of the sensor have different responsivities at the same wavelength and portray different degrees of radiation damage. These properties suggest that the etching or fabrication process is non-uniform across the Sample-B sensor, with variations in the oxide thickness and the interface state density between the pixels. The non-uniformity problem is aggravated after the 157 nm irradiation, where the DUV-induced dark current generation in Sample-B is accompanied by an increase in DSNU and in PRNU.

Since Sample-B has a large PRNU, the visible extrinsic QE is evaluated locally at the individual pixels instead of globally across the sensor. Two sample pixels are considered for the visible QE measurements. One pixel is identified as Pixel-M, and has been exposed to the lower F_2 laser intensity of 0.2 pJ/pulse. The second pixel is identified as Pixel-N, and has been exposed to the higher F_2 laser intensity of 10 pJ/pulse. The results from the visible extrinsic QE measurements are displayed in Fig. 12.14 for Pixel-M and Fig. 12.15 for Pixel-N. For both pixels in Sample-B, the visible QE increases following the 157 nm irradiation, and the increase is more significant for Pixel-N than for Pixel-M. Moreover, the post-DUV measurements reveal the pixels in the B1 sensor, which have been exposed to a lower 157 nm laser intensity, have a better device uniformity and a lower dark current than the B2 sensor.

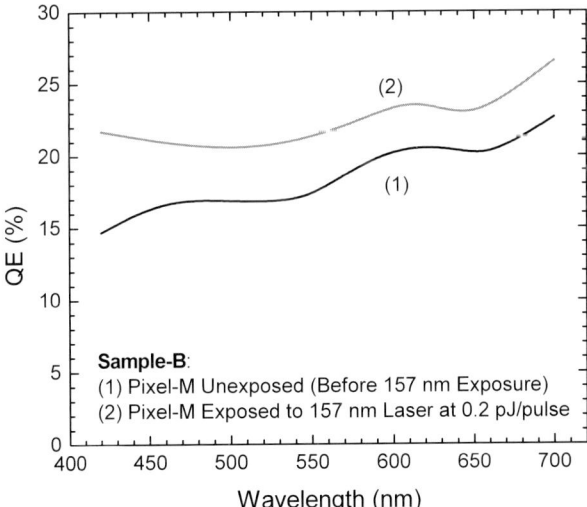

Fig. 12.14. The visible extrinsic QE of a sample pixel in Sample-B, measured before and after the 157 nm exposure at 0.2 pJ/pulse. The measurements are performed at room temperature and at a CCD integration frequency of 15 kHz to 30 kHz

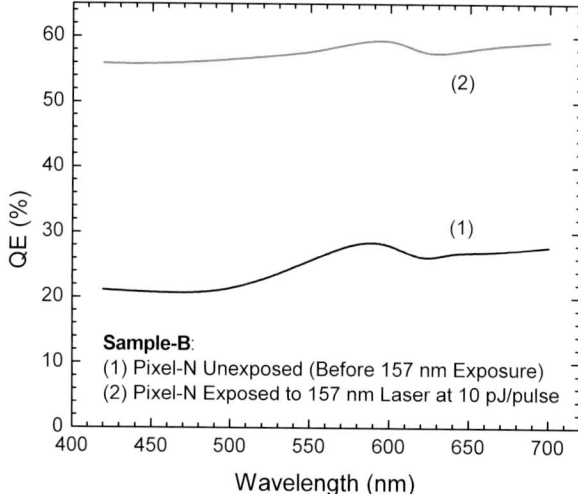

Fig. 12.15. The visible extrinsic QE of a sample pixel in Sample-B, measured before and after the 157 nm exposure at 10 pJ/pulse. The measurements are performed at room temperature and at a CCD integration frequency of 15 kHz to 30 kHz

Contrary to the behavior of Sample-A, the visible QE of Sample-B increases after the 157 nm irradiation. These contrasting observations suggest that the Sample-A and Sample-B sensors are stimulated by different DUV-induced damage mechanisms. One possible explanation requires insight into the relative influence of the DUV-induced effects in SiO_2 (e.g., oxide charge modification) and in the Si-SiO_2 interface (e.g., the creation of interface traps). It appears that the interface state creation is associated with a higher dark current and a lower QE. However, the changes in the QE and dark current, due to DUV-induced effects in the SiO_2, can be quite the opposite.

The research presented in this book provides strong evidence that the DUV-induced color centers and charged defects in the SiO_2 layer can affect the optical and electrical properties of the CCD sensor.[22] If a sufficient number of DUV-induced oxide charges remain after the 157 nm irradiation, they can attract carriers to accumulate at the Si-SiO_2 interface in the Si layer. This can either obstruct or promote interface trapping (an example is shown in Fig. 12.9). If the residual oxide charges reduce the carrier trapping at the interface, this improves the collection efficiency and the visible QE. In addition, the DUV-induced absorption in the SiO_2 can permanently alter the optical characteristics of the CCD. If the transmission of the oxide layer increases or if the reflection decreases following the DUV irradiation, a higher visible extrinsic QE results. Therefore, the DUV-induced effects in the oxide and in the interface can induce contrasting effects on the CCD characteristics. How

---

[22] See Chap. 10 for UV-induced effects in SiO_2.

each of these effects alter the visible QE characteristics of the two sensors before and after the 157 nm irradiation will be examined as four different cases.

Case 1 – Sample-B before DUV Exposure: Since Sample-B has a high intrinsic dark current (i.e., a large number of interface defects), the influence of the interface states most likely outweighs that of the oxide charge defects during the pre-DUV measurements. These interface states can trap electrons and reduce the visible QE of the sensor. This is a potential cause for the smaller QE at visible wavelengths of Sample-B than that of Sample-A before the DUV exposure.

Case 2 – Sample-B after DUV Exposure: The 157 nm irradiation can generate a large density of oxide charges; this is evident from the facts in Sect. 10.5. If it is assumed that a sufficiently large concentration of the DUV-induced oxide charges remain after the DUV exposure, these charges can induce an inversion layer at the Si-SiO$_2$ interface that blocks or passivates all the interface traps. Then, the subsequent carrier generation in the Si layer can contribute to the extrinsic QE of Sample-B. Here, the influence of the UV-induced oxide charges dominates over that of the interface traps, resulting in a net increase in the visible QE after the 157 nm exposure. In addition, the structural properties of the oxide can be modified by the DUV photons in a way that improves the transmission at visible wavelengths. This can also lead to higher visible QE.

Case 3 – Sample-A before DUV Exposure: Sample-A has a thicker oxide and a smaller intrinsic dark current; the latter trait signifies that a smaller number of interface defects exists and have relatively little impact on the QE (compared to Sample-B). Therefore, Sample-A has a higher QE at the visible wavelengths than Sample-B, prior to any DUV irradiation.

Case 4 – Sample-A after DUV Exposure: The interface state density and the fixed oxide charge density can increase after the DUV exposure, as discussed in Sect. 12.3, and in the literature reviewed in Chap. 7 and 8. The discussion in Sect. 10.5 suggested that the amount of increase in the DUV-induced oxide charge density depends strongly on the quality and processing conditions of the SiO$_2$ layer, as well as on the cumulative DUV dose. Since Sample-A has a thicker oxide and is subjected to a different cumulative DUV dose than Sample-B, the DUV-induced oxide charge generation in Sample-A is expected to be different from that in Sample-B. These dissimilarities provide a justification for the contrasting post-DUV behavior in the visible QE observed between Sample-A and Sample-B. If it is assumed that the DUV-induced interface state creation has a stronger influence on the CCD operation of Sample-A than the DUV-induced oxide charge modification, then the dominance of the interface trapping results in a reduction in the visible QE for Sample-A after the 157 nm exposure.

Furthermore, the differences in the QE behavior between the two CCD samples can be attributed to processing-related imperfections. The oxide etching process for Sample-B may have exposed the Si surface to aluminum during a subsequent metal deposition step, causing damage to the interface quality (Sect. 12.1.2). If a thin, residual layer of aluminum is left in arbitrary locations of Sample-B, the residual aluminum can make the sensor susceptible to reflection loss. This is another potential reason for the smaller visible QE in Sample-B, than in Sample-A prior to the DUV exposure. The PRNU of Sample-B can also be linked to the spatial variations in the thickness of the residual aluminum layer. After the 157 nm irradiation, the visible QE is higher for the pixels in Sample-B that have been exposed to the higher intensity F_2 laser irradiation at 10 pJ/pulse. The more intense DUV irradiation may have modified the residual aluminum layer. Consequently, the reflection loss, due to the aluminum, diminishes so that there is an improved visible QE for these pixels.

To validate these assumptions and speculations, it is necessary to conduct a similar sequence of the 157 nm irradiation experiments on simple Si-SiO$_2$ test structures. These free-standing Si-SiO$_2$ test structures are convenient for measuring the interface state density and the oxide charge density. Thus, changes in these device parameters can be carefully monitored as a function of the 157 nm exposure dose to provide an opportunity to confirm the validity of the preceding analyses.

In addition to the analyses and explanations offered in this chapter, there can be alternative interpretations for the CCD behavior observed in the 157 nm irradiation experiment. Also, it is important to note that the measured data are subjected to experimental uncertainties, since some of the measurements are difficult to conduct due to the high PRNU and the high DSNU of the CCD sensors.

12.4.5 DUV-Induced Changes in CCE

The 157 nm irradiation causes the CCE of a Sample-A sensor to decrease from 9.7 µV/e (pre-irradiation) to 8.25 µV/e (post-irradiation). Usually, the CCE is controlled by the CCD sensor's output stage, where the photogenerated charges are transferred across the CCD registers to an output node and are converted to a voltage signal by the output amplifier. Thus, the variation in the CCE can be linked to two main factors: the changes in the gain of the output amplifier stage, and the changes in the capacitance of the output node. With the former, measurements confirm that the output amplifier gain is unaltered after DUV irradiation. For the latter, even though direct capacitance measurement cannot be performed on a packaged CCD sensor, CCE variation due to DUV-induced changes in the output node capacitance appears logical and can be rationalized as follows.

First, it is necessary to recognize that in a typical CCD design, the output node is isolated from the CCD shift registers by a thick field oxide region. If there is UV-induced oxide charging in the field oxide region surrounding the output node diffusion, and if the DUV irradiation generates sufficient

12.5 Response Measurement of Photodiodes at 157 nm

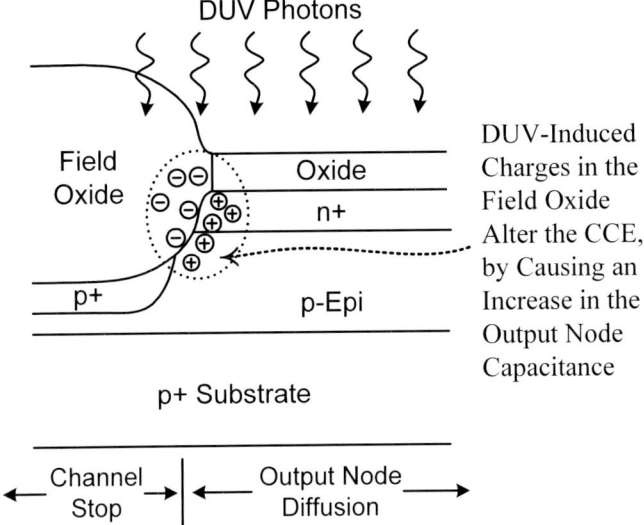

Fig. 12.16. A schematic of the DUV-induced charging in the field oxide to explain the changes in the CCE after the 157 nm irradiation

negative oxide charges in the field oxide, these oxide charges can attract positive carriers in the neighboring Si layer to accumulate at the edges of the diffusion node. This situation is illustrated in Fig. 12.16. Consequently, this charge accumulation leads to an increase in the edge capacitance of the output node. Since the CCE is inversely proportional to the output node capacitance, this increase in the capacitance results in a reduction of the CCE after the 157 nm irradiation. Therefore, the CCE variation can be attributed to DUV-induced changes in the output node capacitance.

The post-irradiation CCE of the Sample-B sensor is difficult to evaluate, because the current and voltage measurements are overwhelmed by the large dark current, large PRNU and large DSNU after the 157 nm exposure. Nevertheless, Sample-B is also susceptible to similar radiation-induced changes in the CCE as the Sample-A sensor.

12.5 Response Measurement of Photodiodes at 157 nm

In addition to irradiating the F_2 excimer laser on the thinned front-illuminated CCD samples, the 157 nm measurement is also conducted on the photodiode (PD) test structures that are fabricated by the same CCD process. With this approach, the behavior of the photosensing element at 157 nm can be pinpointed, while the concerns about the potential DUV damage in the other parts of the CCD sensor (e.g., read-out circuits, CCD registers) are eliminated. Three types of photodiodes are considered in the experiment,

Table 12.3. The types of photodiodes used for the 157 nm response experiment

Device ID	Description
Pinned PD	A regular pinned photodiode, oxide not etched.
Regular PD	A regular n + photodiode, oxide not etched.
Thinned PD	An n + photodiode with thinned oxide overlayers.

as outlined in Table 12.3, which include a pinned photodiode, a regular n + photodiode, and an n + photodiode with thinned overlying oxide.[23]

Figure 12.17 offers the cross-sectional diagrams of the regular n + photodiode and the pinned photodiode, on a p-type Si substrate. The key difference between these two designs is that the latter has an additional p + shallow implant underneath the Si-SiO$_2$ interface. This layer effectively "pins" the potential peak and the photogenerated charge within the middle n-region and away from the Si-SiO$_2$ interface [2]. The potential energy diagram in Fig. 12.18 provides a better visualization of the charge distribution in the two photodiode structures. The additional implant layer in the pinned photodiode helps to control and reduce the interface trapping of the carriers, providing leverage in the performance for the pinned photodiode over the regular photodiode under normal conditions.

These photodiode test structures are exposed to a 157 nm F$_2$ excimer laser irradiation, and their photoresponse are monitored over a two-hour period at laser intensities ranging from 1 nJ/pulse to 4 nJ/pulse. Relatively low laser intensities are selected to avoid saturation of the photodiodes during the response measurement. The experimental setup for the photodiodes is same as that for the CCD sensors, described in Sect. 12.1.1. Figure 12.19 exhibits the normalized response of the photodiode test structures at 157 nm, when irradiation was conducted at a F$_2$ laser intensity of 1.2 nJ/pulse.

The experimental results demonstrate that the thinned photodiode, compared to the non-etched test structures (i.e., the pinned photodiode and the regular photodiode), is the most responsive to the 157 nm laser irradiation. The thinner overlying oxide of the thinned photodiode absorbs less UV photons, and thus leads to an improved response at 157 nm. This observation is in agreement with the data presented in Sect. 12.2 for the thinned front-illuminated CCD sensors, where the CCD with a thinner overlying oxide offers a higher QE at 157 nm than the thicker oxide counterpart.[24] These outcomes further reinforce the fact that reducing the oxide overlayer thickness of the photosensing element enhances the device's responsivity in DUV.

---

[23]The thinned PD is very similar to the photosensing element of thinned front-illuminated CCD sensor samples discussed in Sect. 12.1.2, but with oxide thickness of approximately 60 nm.

[24]Refer to Sect. 12.2.

12.5 Response Measurement of Photodiodes at 157 nm   189

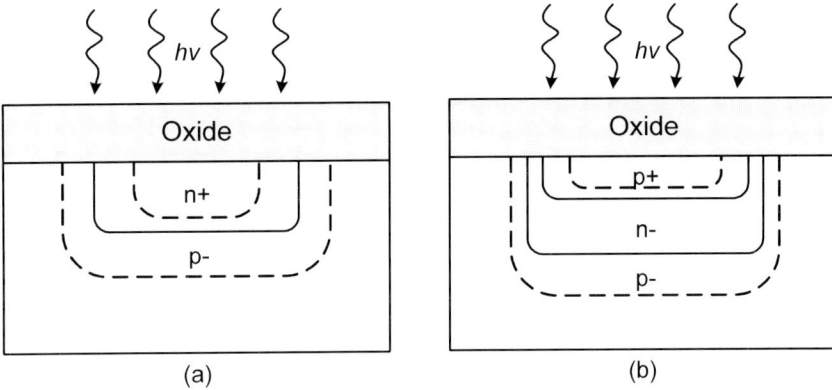

Fig. 12.17. A cross-section diagram of the photodiode test structures: (**a**) a regular n + photodiode on p-Si substrate, and (**b**) a pinned photodiode on p-Si substrate

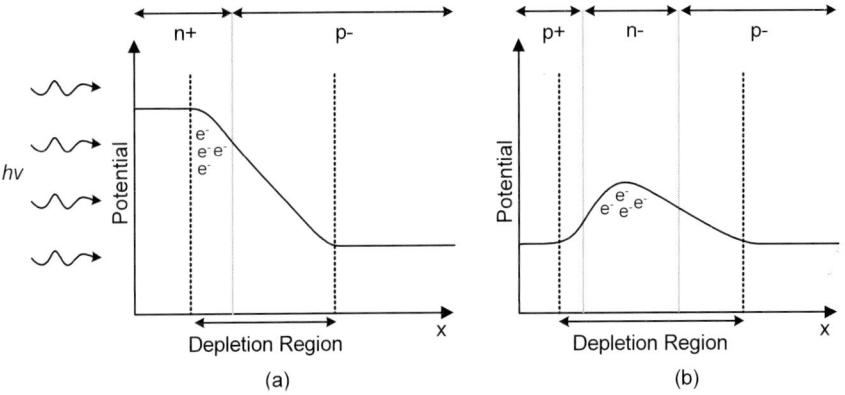

Fig. 12.18. The potential distribution of (**a**) a regular n + photodiode and (**b**) a pinned photodiode

If the responses of the two non-etched test structures are compared, the pinned photodiode exhibits a slightly better QE at 157 nm than the regular n + photodiode. This is attributable to the additional p + implant layer in the pinned photodiode. The p + implant layer not only passivates the interface traps, but it also pushes the potential peak and the photogenerated charges away from the interface. This effectively reduces the carrier trapping at the interface and enhances the efficiency of the pinned photodiode device.

Figure 12.19 illustrates an increase in the 157 nm response of the photodiodes with increasing cumulative DUV exposure. This behavior is similar to that observed in CCD sensors, and is attributed to the UV-induced effects in the SiO_2 and the Si-SiO_2 interface that were in described in Sect. 12.3.1. In summary, the 157 nm experimental results of the photodiode

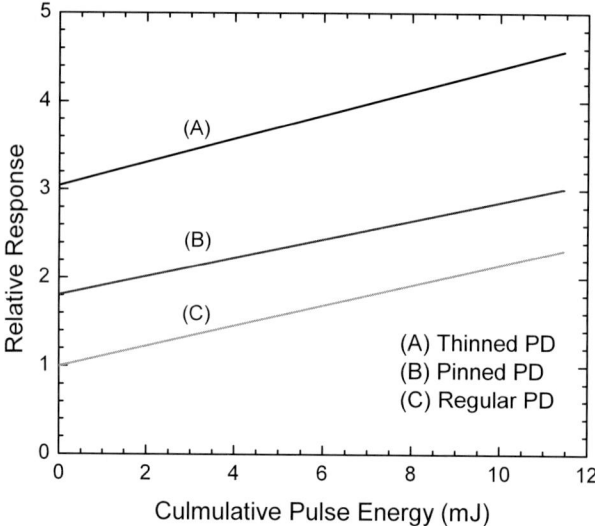

Fig. 12.19. The relative response of the photodiode (PD) test structures at 157 nm. The values are normalized to the initial 157 nm response of the regular PD. The F_2 laser intensity arriving at the sample is estimated to be 1.2 nJ/pulse. The F_2 laser is operated with a pulse width of 15 ns at 10 Hz.

test structures are consistent with the 157 nm response data for the thinned frontside-illuminated CCD sensors. These two sets of data substantiate the validity of the observed response behavior of the CCD sensors at DUV wavelengths, confirming the occurrence of fluctuations in the responsivity or QE of the CCD at 157 nm.

12.6 Summary of CCD Behavior at 157 nm

From the experimental results, it is concluded that the frontside-illuminated linescan CCD sensors with a thinned overlying oxide offer a higher responsivity than the conventional frontside-illuminated CCDs at the DUV wavelength of 157 nm. In addition, the results suggest that the oxide overlayer thickness and the Si-SiO$_2$ interface quality are pivotal factors that determine many of the CCD performance parameters (such as responsivity and long-term stability). The CCDs with the thinner overlying oxide yield a higher extrinsic QE than those with the thicker oxide at the DUV wavelengths. However, a major challenge that researchers are confronted with is the instability of CCD sensors for DUV imaging. QE fluctuations are observed, where the QE of a CCD is enhanced with a continuous 157 nm exposure, but the QE is reduced after the irradiation is interrupted momentarily. Another stability concern is the dramatic increase in the dark current, when a CCD is exposed to a high F_2 laser intensity. In addition, the post-DUV measurements indicate that the visible QE, the dark current, and the CCE of the CCD sensors are permanently altered after the 157 nm F_2 excimer laser irradiation.

The fundamental causes of the DUV-induced degradation of the CCD sensors are very similar to the ionization damage in MOS structures. In both situations, the radiation hardness of the device is determined by the rate of the charge build-up in the insulating layers and by the rate of the interface state generation, as the total radiation dose increases.[25] In addition to the ionization damage effects, the fluctuations in the CCD behavior at 157 nm can be attributed to various UV laser induced effects including the induced absorption in SiO_2, oxide charge modification, interface state generation, and structural modifications. These UV-induced effects alter the electrical and optical properties of the oxide and the interface, leading to a degradation in the device performance. For instance, the UV-induced oxide charges can attract carriers toward the interface in the Si layer, altering the trapping dynamics and the electrical properties of the Si-SiO_2 interface. As a result, the photogenerated signal, the QE, and the dark current of the CCD sensor are affected by this UV-induced oxide charging effect. In addition, the DUV photons have sufficient energy to weaken or break the strained bonds near the interface. This causes an increase in the interface state density and results in the radiation-induced dark current generation in the CCD sensors. Because incident DUV radiation is absorbed close to the Si-SiO_2 interface, careful control of the interface quality with an effort to minimize the interface state density (e.g., the intrinsic interface defects in conjunction with the defect precursors), is critical for improving the DUV performance of the CCD sensors.

In summary, a thinned frontside-illuminated linear CCD sensor that is designed for DUV imaging applications is presented. It is confirmed that the thinned CCD structures are responsive at 157 nm, although further improvements in the device stability are necessary. A careful optimization of the oxide thickness and interface quality are critical to achieve the desired performance, radiation tolerance, and stability at DUV wavelengths for CCD sensors. This investigation demonstrates that the use of thinned front-illuminated CCDs is a feasible and low-cost option for DUV imaging.

12.7 Future Investigations

The DUV-induced fluctuations in the CCD characteristics signify that the properties of the SiO_2 layer and the Si-SiO_2 interface are modified by the F_2 laser photons. For future investigations, the characterization of the oxide and interface properties before, during, and after the DUV irradiation is highly recommended. Such characterization data present valuable information for

---

[25]The ionization damage of MOS devices was presented in Chap. 7.

understanding the degradation behavior and damage mechanisms of CCD image sensors. However, the intricate circuitry and packaging of a CCD sensor inhibit direct measurement or characterization of the the oxide and interface layers.

Instead, simple Si-SiO$_2$ test structures, that are fabricated by the same CCD processing procedures, can provide an effective tool for making the necessary measurements in the oxide layer and at the interface. If these test structures are exposed to 157 nm irradiation, then a sequence of tests can be performed to monitor the changes in the device's properties as a function of the DUV dose. The parameters of interest include the SiO$_2$ absorption characteristics, oxide charge density, defect density, interface state density, and the types of oxide and interface defects. For example, capacitance measurements can be performed on the Si-SiO$_2$ test structures to characterize the interface state density and the oxide charge density. The examination of the DUV-induced effects in the Si-SiO$_2$ test structures may shed some light on the potential degradation processes in a CCD sensor. Such results can also provide a means for validating the analysis on DUV-induced effects in CCD sensors examined in this book. Tracing the origins of DUV-induced degradation in CCD sensors is an essential step for advancing the development of new-and-improved-CCDs for DUV imaging applications.

In addition, it is worthwhile to investigate techniques for recovering the CCD performance after DUV exposure. One approach is the use of thermal annealing, where the irradiated sensor is subjected to high temperature cycles to mitigate the radiation damage. A variety of radiation-induced defects are destroyed at high temperatures. An alternative option is to cool the CCD sensors. Since the dark current has a strong dependence on the temperature, it is possible to effectively suppress or eliminate the DUV-induced dark current by cooling the CCD sensor to very low temperatures. Nevertheless, an indepth investigation is necessary to determine the optimal process parameters for the thermal annealing or the cooling of CCDs that are targeted for DUV applications.

Part VI

Concluding Remarks & Future Research

13 Design Optimizations for Future Research

The analyses in the preceding chapters have indicated that the DUV-induced damages in CCD sensors are primarily manifested as defects or structural modifications in the oxide layer and the Si-SiO$_2$ interface. Thus, it is reasonable to postulate that the suppression of defects in the oxide and interface presents a feasible approach for managing radiation damage in CCDs. Theoretically, if the oxide thickness is reduced, the extent of the photon-oxide interaction decreases also. Furthermore, if the device has a high-quality Si-SiO$_2$ interface (i.e., very few strained bonds, a low interface state density, and a small degree of structural disorder in the network), the generation of radiation-induced interface states is mitigated, since there are very few precursors of dangling bonds or interface traps. Therefore, it appears that oxide thinning and the control of the interface quality are viable techniques for improving the tolerance of CCD sensors to DUV radiation.

Yet, how thin should the oxide layer be? The expectation may be that the thinner the oxide, the better the device performance and the radiation hardness. However, studies on UV Si photodiodes have showed that there is a minimum oxide thickness beyond which the device performance degrades significantly with increased UV exposure. It has also been found that the nitridation of the Si-SiO$_2$ interface enhances the UV stability of photodiodes [67]. Nitridation facilitates a trap-free interface, which is critical for photodiodes and CCDs alike. The process helps to eliminate interface recombination at short wavelengths, where the photogenerated carriers are created in the vicinity of the interface. The investigation of UV photodiodes will be followed up in Sect. 13.1.

In addition to the oxide thickness, the quality and doping of the Si and SiO$_2$ affect the radiation hardness of CCDs. For instance, the proper control of the doping density and dopant diffusion in the Si layer can improve the carrier generation and collection processes, and minimize the recombination loss of the carriers in the Si layer and at the interface. The research on DUV silica glass provides additional insight into optimizing CCD sensors for imaging at DUV wavelengths. The semiconductor industry continues to pursue the development of silica glasses with a high transparency and long-term stability for DUV lithography applications. Researchers have found that fluorine doping improves the stability of SiO$_2$ glasses, where the incorporated fluorine atoms reduce the strain and disorder in a SiO_2 network [58]. The fluorine doping of SiO_2 will be examined in Sect. 13.2, and the viability of using fluorine doping for extending the lifetime and DUV stability of a CCD sensor will be addressed.

The optimization methods in this chapter are derived from the research results that are related to DUV imaging, and offer a starting point for the future optimization of DUV-sensitive CCD sensors. It is expedient to implement these novel techniques in the design of CCDs, provided that the techniques are compatible with the existing CCD fabrication process. Foremost, it is essential to examine the compliance of these new design techniques with the constraints and objectives of the CCD design, and the manufacturing process (e.g., technological compatibility and cost requirements). For the frontside-thinned front-illuminated CCD sensor that is presented in this book, the main attraction is its lower production cost and easier implementation than that of the other specialized UV-sensitive CCD sensors (e.g., backside-thinned back-illuminated CCD). Consequently, the incorporation of the proposed optimization techniques in thinned frontside-illuminated CCDs are justified only if key features such as cost efficiency and ease of implementation can be accommodated.

13.1 Optimization Techniques Based on UV Photodiodes

Because incident UV radiation is absorbed by Si within a few tens of nanometers of the $Si-SiO_2$ interface, efforts to minimize the recombination at and near this interface is crucial for attaining a Si-based solid state detector (e.g., CCD sensor and photodiode) with a high responsivity and long term stability in UV. Moreover, UV radiation generates oxide-trapped charges which contribute an electric field perpendicular to the interface. The resulting electric field attracts the photogenerated carriers in the Si layer towards the $Si-SiO_2$ interface, altering the electrical characteristics and stability of the device. As a result, the key criteria for achieving a stable and sensitive UV detector are to minimize the interface state density and to suppress oxide charge generation. The techniques and strategies, contrived by the Si photodiode industry to address these UV issues, will be reviewed next.

13.1.1 Characteristics of Photodiodes in UV

The utilization of semiconductor photodiodes, originally designed as visible-light detectors, is extended to the UV and X-ray regions. However, these photodiodes often exhibit a poor efficiency at these wavelengths, owing to the strong absorption and reflection of the surface materials [69]. In addition, the underlying semiconductor substrate has high absorption coefficients

13.1 Optimization Techniques Based on UV Photodiodes

at these wavelengths; thus, the photons are absorbed in the semiconductor very close to the interface. The shallow photon absorption depth is a problem because the large e-h recombination rate at this interface can cause further degradation in the efficiency of the photodiode. The reaction of the photodiode material with the absorbed highly energetic photons is also a relevant concern. In the case of SiO_2, ionization damage can be stimulated by UV and X-ray photons. Alike most other Si-based devices, the SiO_2 layer has a dominant influence on the performance parameters and radiation hardness of the UV photodiodes. A stable and efficient UV photodiode can be fabricated with a precise control of the SiO_2 layer and the Si-SiO_2 interface quality. The discussions in the preceding chapters have shown that similar challenges apply to CCD sensors in DUV.

Research groups have focused on the development of Si photodiodes for UV and X-ray detection, and have identified three main obstacles in fabricating a good Si XUV (the abbreviation for soft X-ray and vacuum-UV) detector: the presence of a dead region at the photodiode surface (formed during the fabrication process), a relatively poor quality of surface passivation SiO_2, and an instability in the QE [70]. Due to the extremely short absorption depths of the XUV photons in the Si, the presence of a dead region at or near the diode surface absorbs most of the photogenerated carriers; this carrier loss translates into an unacceptably low QE. The quality of the surface passivation oxide is critically important as well, because this layer is, effectively, an entrance window for the incoming photons, and any interaction with XUV photons tends to degrade the device efficiency. In addition, conventional Si photodiodes exhibit an unstable QE with UV irradiation. The QE instability in photodiodes is believed to be a consequence of the increasing surface recombination at the radiation-induced interface traps. This argument is substantiated by the QE degradation measurements of the UV photodiodes which demonstrate that the interface trap area density, N_{it}, increases with the increase in the UV dose [71]. Similar QE instability was reported for CCD sensors in the DUV in Chap. 12, and it was concluded that interface state creation is one of the mechanisms responsible for the QE fluctuation.

Other potential causes for the loss of the short-wavelength QE in Si photodiodes are also proposed, including unsatisfactory Si-SiO_2 interfaces, latent recombination centers near the interface, moisture absorption by the device, and dopant distribution. Unsatisfactory interfaces and latent interface traps are the consequences of inferior interface structures derived from bulk traps in the vicinity of the interface, the existence of strained bonds at the interface, the presence of excess H and OH in the oxide, excessive boron in the oxide, and so on. Moisture, in conjunction with the boron impurity, is also suspected of generating recombination centers near the Si-SiO_2 interface. All of these imperfections must be minimized to realize a photodiode with a stable QE in UV [67]. Furthermore, since UV photons are absorbed in the first few hundred nanometers of Si, the boron and phosphorous diffusions in the Si must be shallow and well-controlled to minimize the carrier recombination between the Si-SiO$_2$ interface and the p-n junction of the photodiode.

13.1.2 Techniques to Improve the UV Performance

To remedy the problem of the QE instability in Si photodiodes, Korde et al. have investigated three design strategies for the fabrication of UV-enhanced Si photodiodes: a high quality SiO$_2$ coating, an optimized SiO$_2$ thickness, and a nitridation of the passivating SiO$_2$ layer [70]. The careful integration of these features yields an extremely stable n-on-p diffused UV-enhanced Si photodiode with practically no dead region at the surface and a passivating SiO$_2$ coating that is optimized for the UV regime. Also, this design has no carrier recombination at the Si-SiO$_2$ interface or in the front diffused (doped n+) region. Thus, the design exhibits a near-theoretical QE at UV to soft X-ray wavelengths [70]. These optimization techniques are examined in greater detail in this section.

13.1.2.1 SiO$_2$ Quality

Often, the SiO$_2$ layer has multiple functions in Si-based devices; namely, the antireflection coating, interface passivation layer, and gate insulator. Irrespective of its exact functionality, the oxide layer is typically the entrance window for the incident photons. Thus, the oxide properties must be controlled and optimized to ensure a minimal photon loss. Studies have revealed that one of the criteria for a stable UV photodiode is that the final oxide needs to be a MOS grade SiO$_2$ coating, grown by using a clean, dry oxide process, and a proper annealing step is required to ensure device stability during the UV exposure [72].

Korde et al. have established that the triple-wall oxidation (TWO) furnace system offers a clean and dry oxidation process that is appropriate for attaining the desired SiO$_2$ quality for UV Si photodiodes. In comparison, the standard method, using a single-wall oxidation (SWO) furnace process, is inferior for the UV photodiode fabrication. The conventional SWO method is at stake, since it introduces defects or impurity traps in the oxide by allowing the diffusion of heavy metal ions and mobile alkali ions through the quartz tube wall. Also, a considerable amount of moisture is incorporated into the oxide via the diffusion of water through the single quartz wall. These adverse conditions, seen in the SWO process, are avoided in the TWO furnace system [73]. It is demonstrated that the thin gate oxide, grown by using a TWO furnace system, exhibits an improved dielectric breakdown field strength with less interface trap generation than the gate oxide that is grown in a conventional SWO system. Along with an enhanced SiO$_2$ quality, microscopic as well as macroscopic defects are better controlled in a TWO system. Korde et al. have successfully incorporated a TWO process into the fabrication of extremely stable UV-enhanced Si photodiodes, and the resulting devices are characterized by an excellent radiation hardness and a 100% internal QE [71]. For more details on the TWO system, consult the reference for Yoon et al. [73].

The supervision of the Si-SiO$_2$ interface quality and trap density is also important during oxidation. To grow reliable thin oxides, the Si-SiO$_2$ interface must be first prepared with a low concentration of hydrogenated dangling bonds. A high density of hydrogenated dangling bonds can trigger electrical instabilities in Si-based devices. Hydrogen bonds at this interface can be broken by a hot electron or an ionizing radiation stimulation, where active or charged electronic traps are generated from these dangling bonds. Another requisite for the growth of a reliable oxide is a low active trap density so that even if the traps are activated, the number of trapped charges is too small to disturb the device characteristics. These requirements are satisfied by the TWO process. Since it facilitates a better control of the water-related defects or hydrogenated dangling bonds, the resulting interface has a lower concentration of dangling bonds and a lower active trap density [73].

13.1.2.2 Optimal Oxide Thickness

At photon energies greater than 10 eV approximately, significant absorption can occur in the SiO$_2$ layer [67]. To lessen the photon-oxide interaction, the thickness of the SiO$_2$ layer on the photodiode needs to be reduced. But what is an appropriate SiO$_2$ thickness for UV detectors? Canfield et al. have evaluated the QE stability and the spatial uniformity of UV Si photodiodes with different oxide thicknesses [67]. A cross-section of the test structures of Canfield's experiment is given in Fig. 13.1, which are n-on-p photodiodes fabricated on Si (111) wafers, with a phosphorus or arsenic diffusion. The thermally grown oxide layers are etched to thicknesses in the 20 Å to 250 Å

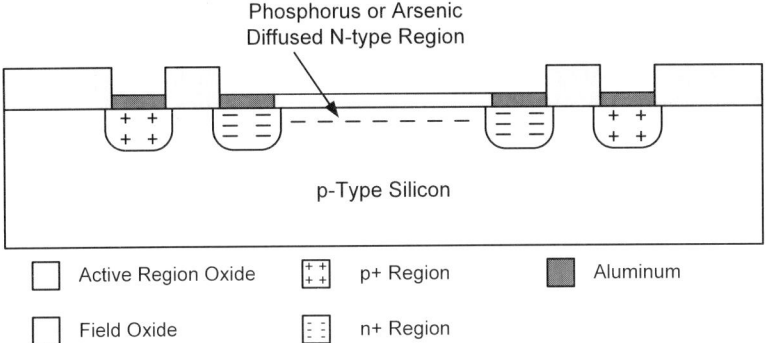

Fig. 13.1. A cross-section of a commerical (n-on-p) UV photodiode (adapted from Canfield et al. [67])

range. The photodiode samples have a 1 cm² circular active area, and are operated in a windowless fashion.

The experimental results provide evidence that the QE and stability of Si photodiodes in the UV and X-ray regions are significantly influenced by the SiO$_2$ thickness. Figure 13.2 presents the QE of photodiodes with different oxide thicknesses as a function of the photon energy [67]. As the incident photon energy increases, the QE improves with the decrease in the oxide thickness. Moreover, the photodiode with the thinnest oxide achieves the highest QE at photon energies near 10 eV [67]. This is logical because the maximum absorption of SiO$_2$ occurs near 8 eV to 10 eV, and thus the impact of oxide thickness on the QE is more pronounced at these photon energies.

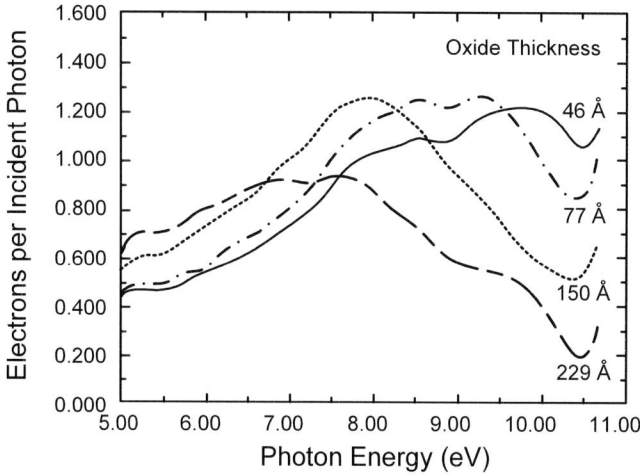

Fig. 13.2. The QE of UV photodiodes with different oxide thicknesses, for photon energies in the region of oxide absorption (adapted from Canfield et al. [67])

In addition to evaluating the efficiency of the XUV photodiodes, Canfield et al. have examined the stability of XUV photodiodes by measuring the QE at regular intervals during a long-term exposure to XUV radiation. A photodiode with 44 Å of oxide is tested and its stability is assessed at three photon energies (7.7 eV, 10.2 eV, and 124 eV) for an exposure of up to 3.0×10^{14} photons/cm²; the results are summarized in Fig. 13.3 [67]. Of the three wavelengths, the 10.2 eV ($\lambda = 121$ nm) radiation induces the largest QE degradation because significant oxide absorption occurs at this energy. In contrast, the QE at photon energies of 7.7 eV and 124 eV remain approximately constant (with minor fluctuations) as a function of the cumulative exposure dose. These results confirm that the oxide absorption, particularly in the UV-VUV region where the absorption is maximum, indeed imposes a serious concern on the stability of Si-based imaging devices.

13.1 Optimization Techniques Based on UV Photodiodes

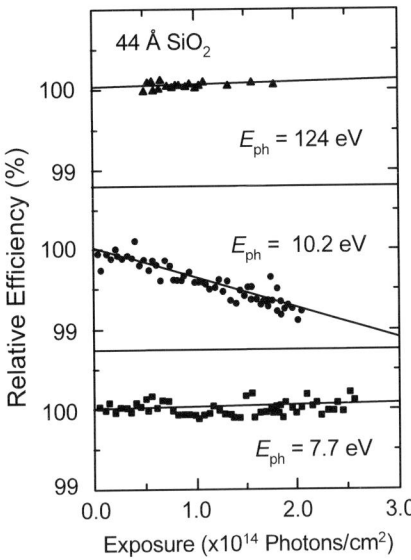

Fig. 13.3. The effect of the XUV radiation on the stability of a UV photodiode with 44 Å of oxide, at photon energies of 7.7 eV, 10.2 eV and 124 eV (adapted from Canfield et al. [67])

Although Fig. 13.2 reveals that a thinner oxide typically yields a better device performance, insufficient oxide can have detrimental effects. This is apparent in Fig. 13.4, where the QE of a photodiode with 28 Å of oxide is inferior to that of a photodiode with 46 Å of oxide [67]. One potential explanation for the inferior performance of the photodiode with 28 Å oxide is that there is insufficient oxide to properly passivate the Si surface; as such, the Si-SiO$_2$ interface region is characterized by a high density of defects and traps. These defects and traps can aggravate the recombination loss at the interface and in the Si substrate, which subsequently degrade the photodiode's efficiency.

Figure 13.5 exhibits the effect of prolonged radiation exposure (at 10.2 eV) on a photodiode with insufficient oxide of 28 Å thick [67]. The measured QE increases with exposure, but the QE drops when radiation is resumed after it had been turned off momentarily. Similar fluctuations in efficiency are observed when the photodiode is irradiated at photon energies with little oxide absorption. This suggests that a photodiode with insufficient oxide generally has poor stability in UV. An important conclusion, based on these results, is that a thinner oxide usually yields a higher QE, but there is a critical thickness for the passivating oxide layer below which the device performance is seriously degraded. It is evident that the QE fluctuations of the photodiode in Fig. 13.5 resemble the QE fluctuations of CCD sensors with 157 nm irradiation (see Fig. 12.4). This resemblance confirms that the

202   13 Design Optimizations for Future Research

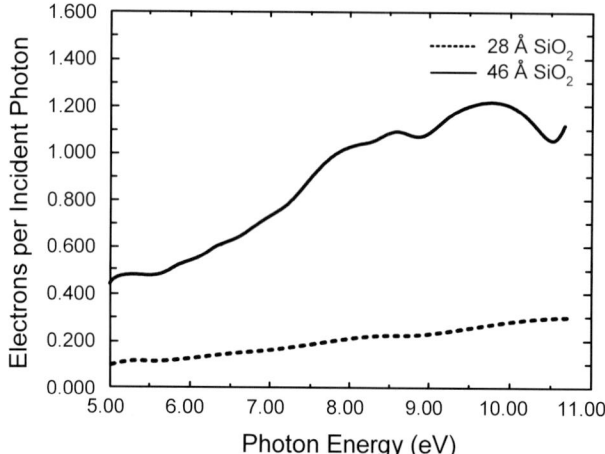

Fig. 13.4. The measured QE of UV photodiodes with 46 Å and 28 Å of oxide. The poorer QE of the photodiode with 28 Å of oxide suggests that insufficient oxide can have detrimental effects on the device performance (adapted from Canfield et al. [67])

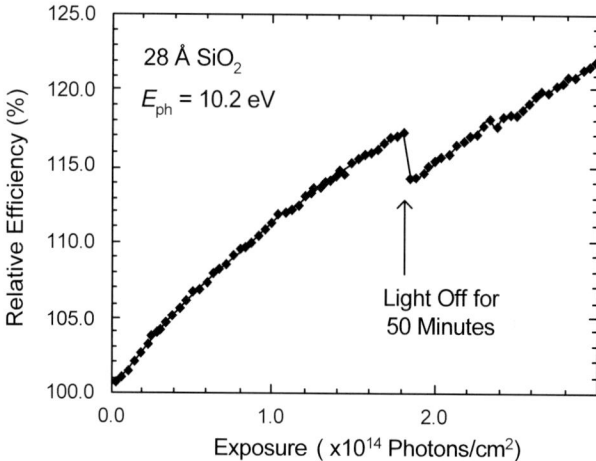

Fig. 13.5. The measured effect of radiation exposure on a photodiode with insufficient (28 Å) oxide (adapted from Canfield et al. [67])

13.1 Optimization Techniques Based on UV Photodiodes

observed fluctuations are not due to experimental errors, and that similar radiation damage mechanisms have emerged to alter the material and device properties of both photodiodes and CCDs.

The investigation demonstrates that the optical absorption in the oxide layer degrades the efficiency of the photodiodes in UV, and it is possible to lessen the absorption-induced-disturbance by reducing the oxide thickness. Conceptually, it is expected that the photons, absorbed in the oxide layer, will not make a positive contribution to the device efficiency. However, Canfield et al. have offered proof that there can be some electronic contribution from a portion of the photons that are absorbed in the oxide. They have compared the measured QE with the calculated QE (assuming there is no oxide contribution) of a photodiode with 77 Å of oxide, and the data are displayed in Fig. 13.6 [67]. The plot indicates that the measured QE is greater than the calculated/theoretical QE (solid curve), over an energy range where the oxide absorption is significant (i.e., near 10 eV). When the theoretical QE is re-calculated with the assumption that 20% of the photons, absorbed in the oxide, contribute carriers to the Si, the modified curve (dashed line) coincides with the experimental QE. This signifies that a percentage of the absorption events in the oxide layer produce carriers that add to the carrier collection in the Si layer, resulting in a positive contribution to the QE of the photodiode. Perhaps, the photogenerated electrons in the oxide drift to the Si layer and supplement the carrier collection. This behavior appears to

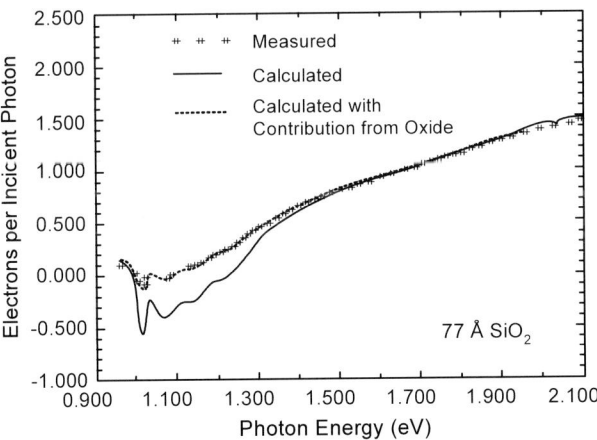

Fig. 13.6. The theoretical and measured QE of a UV photodiode with 77 Å of oxide. The solid curve is QE calculated, based on an efficiency of 3.63 eV per pair for an e-h production in silicon, and includes the losses due to reflection and oxide absorption. The dashed curve is calculated, based on the additional assumption that 20% of the photons absorbed in the oxide contribute carriers to the silicon (adapted from Canfield et al. [67])

play a critical role in the QE fluctuations of CCDs at 157 nm, as stated in Sect. 12.3.

From these results, it can be concluded that careful control of the quality and thickness of the SiO_2 layer, as well as the quality of the Si-SiO_2 interface, is crucial for the realization of stable and efficient UV photodiodes. Also, it is important to recognize that the reduction of the SiO_2 thickness is limited to a critical threshold, below which the device behavior is unacceptable. To achieve radiation-hard UV photodiodes, the oxide thickness should be optimized for the specific wavelength region of interest. For instance, the photodiodes from International Radiation Detector Inc. (IRD), designed for detection in the longer UV wavelength region, feature oxide thicknesses of 40 nm to 170 nm. The photodiodes for UV and X-ray detection feature extremely thin SiO_2 with thickness between 3 nm and 7 nm [74]. Therefore, the oxide thickness of the imager must be prudently selected according to the type of radiation for the target application.

13.1.2.3 Nitridation

The incorporation of a nitridation process during the oxide growth offers another approach for optimizing the performance of Si photodiodes in UV. Nitrided devices show improvements in at least three areas [71]. First, Si photodiodes with nitrided passivating SiO_2 coatings are several times more resistant to ionization damage than diodes with thermally grown oxide. Secondly, a photodiode, nitrided in pure N_2O at 1065°C, has no surface dead region, indicating that there is no recombination of the photogenerated carriers in the doped n+ region or at the Si-SiO_2 interface. The reduced recombination loss enhances the device efficiency. Lastly, the incorporation of nitrogen at the Si-SiO_2 interface renders the device practically insensitive to moisture. For example, the QE (at $\lambda = 50$ nm to 250 nm) of a nitrided-oxide photodiode does not change after an exposure to 100% relative humidity for four weeks [71].

The improvement in the radiation hardness for photodiodes with an increasing degree of nitridation can be explained by the Bond Strain Gradient model. According to this model, interface states are caused by the migration of the radiation-induced defects within the strained region toward the Si-SiO_2 interface, with the strain acting as a driving force. Thus, the amount of interface state generation depends on the degree of strain at the interfacial region, and on the distance in the oxide over which radiation-induced defects undergo strain-induced migration. With nitridation, the presence of nitrogen relieves some of the strain in the interfacial region [71]. Thus, as the nitridation increases, the number of strained bonds at the interface and the width of the strained interface region are reduced. A smaller interface state density is equated to a higher radiation hardness of the Si photodiode.

The effectiveness of nitridation depends on the processing conditions, including the temperature and the types of reactants/chemicals used during the nitridation. For example, a nitridation temperature of 1065°C yields photodiodes with a higher radiation hardness those obtained with nitridation at 1165°C [71]. Figure 13.7 compares the efficiency of three photodiodes with different SiO_2 coatings (each is subjected to different nitridation conditions) upon exposure to XUV radiation. Of the three diodes, the diode whose SiO_2 is nitrided in pure N_2O at 1065°C with a nitrided oxide thickness of 5 nm is the most stable. The second diode, nitrided in ammonia (NH_3) and re-oxidized, exhibits a lower stability than that of the diode nitrided in N_2O. The diode without nitridation shows the fastest decay of efficiency as a function of the UV dose, implying that the diode is the least stable. It can be concluded that the nitridation of the SiO_2 layer improves the radiation hardness of Si photodiodes to ionizing radiation; and more specifically, nitridation using pure N_2O produces photodiodes with better radiation hardness compared to nitridation using NH_3.

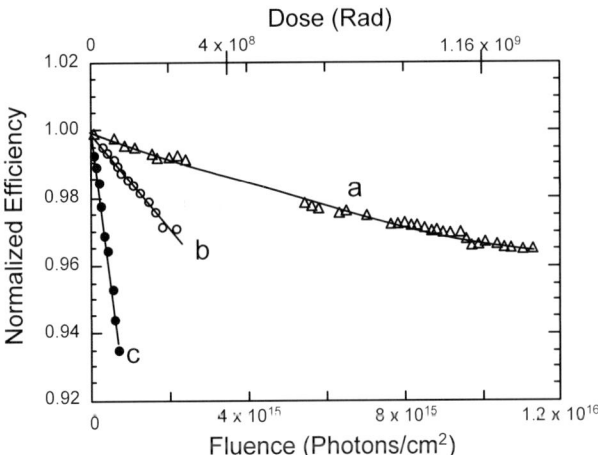

Fig. 13.7. The efficiency of three silicon photodiodes as a function of the fluence (or equivalent dose) at 10.2 eV photon energy. The photodiode (**a**) with nitrous oxide nitridation, (**b**) with ammonia nitridation, and (**c**) without nitridation. The experimental data points are shown, with a least-squares fitted linear (**b** and **c**) and polynomial (**a**) curves (adapted from Korde et al. [71])

In summary, to obtain photodiodes with the desired performance and stability in the UV region, the following techniques are incorporated: the elimination of the dead region, the use of a proper oxidation procedure (e.g., TWO process) to improve the oxide quality, the optimization of the minimal oxide thickness, and the nitridation of the oxide to improve the oxide and interface quality. With these design strategies, Si photodiodes can be fabricated with a high QE in UV. Nevertheless, there is still a persisting challenge to maintain a high QE and good stability in the UV over periods of years.

On-going investigations are being conducted to resolve this UV-induced instability problem in Si detectors. The techniques that are presented here can be incorporated into the CCD design to strengthen the efficiency and the radiation tolerance of CCDs to DUV irradiation.

13.2 Optimization Techniques Based on DUV Silica Glass

Chapter 10 reported that the DUV-induced absorption in SiO_2, due to color center formation, is one of the major consequences of DUV damage. Since the strained Si-O-Si bonds in the SiO_2 network structure control the VUV absorption tail, as well as some of the DUV-induced defect formation processes, the suppression of the strained bonds is a feasible approach to improve the DUV properties of SiO_2. The experiments proved that fluorine doping into SiO_2 glasses effectively enhances both the transmittance at 7.9 eV and the radiation toughness to F_2 laser irradiation. Fluorine doping helps by narrowing the Si-O-Si bond angle distribution (or equivalently reducing the fraction of strained bonds in SiO_2). The effect of fluorine doping on the distribution of strained bonds in SiO_2 is displayed in Fig. 10.6. As the fluorine atom is incorporated into the SiO_2 material, the formation of Si-F bonds breaks the continuity of the SiO_2 network structure at the locations of high strain. This decreases the number of three- and four-membered ring structures (i.e., strained bonds) which, otherwise, contribute to a higher degree of disorder and a poorer transmission at 7.9 eV [58]. As a result, the narrowing of the Si-O-Si bond angle distribution via fluorine doping can suppress the defect formation and enhance the VUV transparency of SiO_2. The idea of fluorine doping of SiO_2 can be integrated into the CCD process to improve the reliability of CCD sensors for DUV imaging applications.

14 Concluding Remarks

14.1 Conclusions

The growing popularity of DUV excimer lasers in industrial applications demands DUV-sensitive imaging systems for detection, inspection, and process control purposes. A DUV-sensitive CCD camera is an ideal candidate for these purposes owing to the camera's high-speed, low-noise, and digital imaging capabilities. A thinned frontside-illuminated linear CCD image sensor, designed for DUV imaging applications, is introduced in this book. The experiment demonstrates that these thinned CCDs are responsive to DUV radiation. However, further improvements in the CCD's stability and radiation tolerance are necessary. It is found that the oxide overlayer thickness and the Si-SiO$_2$ interface quality have a paramount impact on many of the CCD performance parameters in DUV. CCDs with a thinner overlying oxide exhibit a higher QE, and CCDs with an inferior interface quality undergo a more severe dark current instability. The two primary radiation damage effects, observed in the CCDs due to the 157 nm F$_2$ excimer laser irradiation, are the fluctuations in the extrinsic QE and the dramatic increase in the dark current as a function of the DUV exposure dose. The visible QE, the dark current, and the CCE are also permanently altered after the F$_2$ laser irradiation. A detailed analysis of the mechanisms that are responsible for the DUV-induced degradation and instability of CCD image sensors is presented in this book for the first time.

Although it is difficult to precisely state the underlying causes of the CCD instability in the DUV due to the limited experimental data available, potential mechanisms for the CCD degradation are proposed here and are theorized, based on publications in the scientific research arena. The existing literature provides the necessary background on the properties of a Si-SiO$_2$ system and the fundamentals of the material interaction with radiation. This information establishes a foundation for understanding and predicting reactions that can occur in a CCD sensor, when it is subjected to ionizing DUV irradiation. In addition, the studies of silica glasses for DUV lithography applications provide a framework for explaining the DUV laser induced damages in the SiO$_2$ layer of CCDs. The findings from these research fields facilitate an analysis of the origins of the CCD instability in DUV. Specifically, it is recognized that the DUV induced damage on CCD sensors is very sensitive to the SiO$_2$ composition, since SiO$_2$ absorbs DUV laser photons to result in photoreactions that affect the CCD performance. The dominant photoreactions include the DUV ionization damage effects and the DUV laser radiation effects such as the UV-induced oxide absorption due to the color center formation, the generation of oxide charges and traps, the creation of interface states, and the bond and structural rearrangements. These DUV-induced effects alter the electrical and optical properties of the SiO$_2$ and the Si-SiO$_2$ interface layers, and bring forth anomalous CCD behavior. Moreover, because some of these effects are dose-dependent, they provoke temporal and permanent shifts in the CCD characteristics that are responsible for the DUV-induced QE and dark current fluctuations.

Further optimization of the oxide thickness and the oxide and interface quality is necessary to develop CCD sensors with the desired performance, radiation tolerance, and stability at DUV wavelengths. The investigation and analysis presented here will hopefully establish a basis for the future development of new-and-improved DUV-sensitive CCD image sensors. In conclusion, the results demonstrate that the thinned frontside-illuminated linear CCD sensor is a feasible low-cost alternative for DUV imaging applications. Due to the complexity of the mechanisms involved and given the importance of DUV detection in photolithography and sensing applications, more in-depth investigations are essential to quantify the DUV laser interactions with the CCD's underlying material systems. Research on DUV enhancement and radiation hardening techniques for CCD sensors is also crucial in order to drive additional advancements in this technology.

14.2 Recommendations

In the future, it is recommended that researchers focus on perfecting and optimizing the CCD manufacturing process for the fabrication of radiation-hardened DUV-sensitive sensors, conducting more thorough and exhaustive DUV experiments, and investigating the recovery and annealing techniques for CCD sensors after DUV irradiation. Detailed descriptions of these recommendations are provided next.

1. For the thinned frontside-illuminated CCD sensor, the etching process that is currently employed for thinning the oxide on top of the imaging region needs to be refined in order to acquire a stricter control of the parameters, including the oxide thickness, oxide quality, and interface state density. The revised procedure must also ensure minimal processing-induced or impurities-induced damage at the Si-SiO$_2$ interface. By reducing the initial defect concentration in the oxide and the interface, significant improvement in the radiation hardness of CCD sensors at DUV wavelengths can be expected.

2. Once the current manufacturing issues are resolved, the optimization techniques of Chap. 13 can be investigated to achieve additional enhancements on the overall performance and radiation tolerance of the CCD sensors with DUV irradiation. The cited design techniques include the optimization of the SiO_2 thickness and quality, the growth of SiO_2 using a triple-wall oxidation (TWO) system, the nitridation of the SiO_2 layer and the Si-SiO_2 interface, and the fluorine doping of SiO_2. These optimization strategies should be adopted in the CCD fabrication process, provided that they satisfy the various cost and technological constraints of the CCD design cycle. The investigation of new hardening techniques to promote CCD tolerance to DUV radiation is also recommended.
3. An exhaustive and systematic DUV radiation experiment can be conducted to obtain more comprehensive data on the characteristics of CCD sensors, and their dependence on the various DUV exposure parameters. The measurements should be performed in a well-controlled radiation environment, which entails an accurate provision for the parameters such as the radiation intensity and dose, and the device and ambient temperatures. For instance, when the changes in the radiation-induced dark current are monitored throughout the irradiation period, careful control of the device temperature and the ambient temperature is essential because the thermal effects (e.g., laser-induced heating of the sample) can contribute to dark current generation. In addition, it is desirable to conduct measurements on a variety of CCD samples with a wider range of SiO_2 thicknesses, to facilitate the determination of the optimal oxide thickness of CCDs for DUV imaging.
4. The characterization and testing of simple Si-SiO_2 test structures at DUV wavelengths can provide valuable information on the instabilities of CCDs and verify the DUV damage mechanisms proposed in this book. Of prime importance are the measurements of the changes in the interface state density, the oxide defect density, the oxide charge density, and the induced SiO_2 absorption intensity, as a function of the DUV intensity and dose. These test structures should be constructed by using the CCD fabrication process, as such, the data obtained for the Si-SiO_2 test structures can be conveniently correlated with the DUV data for CCD sensors. This will provide more insight into how the interface state density and the oxide charge density of the CCD sensor vary with the DUV dose. It is also favorable to determine the exact nature and the characteristics of the oxide and interface defects that are introduced by DUV irradiation. For example, information on the energy level associated with the defect, trapping and detrapping characteristics of the defect, and the type and polarity of the trap can contribute to the development of a comprehensive model for the DUV-induced degradation processes in CCD sensors. Moreover, the results of these experiments can assist in estimating the magnitude of contributions of the various mechanisms that were examined in Sect. 12.3.

5. It is advantageous to investigate potential recovery techniques to suppress the DUV-induced dark current and to recover the CCD performance after DUV irradiation. For example, thermal annealing is commonly used in the semiconductor industry to recover electronic devices from radiation damage. Another option is the cooling of the CCD sensors to help suppress or eliminate radiation-induced dark current; this is typically performed in a liquid nitrogen dewar. Alternative recovery approaches should also be explored.

Glossary and Definition of Acronyms

Absorption Band: The area of the spectrum in which the absorption coefficient is at maximum.
Absorption Coefficient: The internal absorbance of a material; given that the unit transmission of a material is t, the absorption coefficient is $a = \log_e t$.
Absorption Length: The depth to which 63% ($= 1 - e^{-1}$) of the photons are absorbed by the material at a specified wavelength.
Absorption: The interaction that takes place between the optical electric field of light and the electrons of a solid or gas material; in quantum mechanics, absorption is the interaction between photons and electrons in the presence of the lattice of nuclei.
A-Center: Also called an oxygen vacancy or O-V center; a trap that is created when a silicon vacancy interacts with an oxygen atom to form a stable trap defect.
Amorphous: A disordered, glassy solid state of a semiconductor (or other substance). Amorphous material does not have long range order; the thermal, optical, electrical, physical, and other properties of amorphous material vary considerably in comparison to the highly ordered crystalline state of the same material.
Amplifier Sensitivity: The sensitivity of the output amplifier from electrons to volts (V/e^-).
Angstrom (Å): A unit of length commonly used in semiconductor nomenclature; 1 Å equals 10^{-10} meters.
Annealing: A process wherein a solid material is heated and cooled in order to change its physical, optical, electrical, or thermal properties. UV laser annealing is performed to reverse the damage of the crystal lattices of the semiconductor layers from high energy ion bombardment (ion implantation).
Antireflection (AR) Coatings: A coating that is employed to reduce reflection loss; commonly used in backside-illuminated CCDs.
Aperture: An on-axis, light-restricting mask or object in an optical system; a circular, square, or other (polygonal) shaped physical object that blocks radiation in an optical system from the object side.
Attenuation: A deliberate reduction of the energy or light, generally in an optical system by the insertion of an on-axis element that reflects a portion of the beam from the optical path; attenuation or energy loss occurs (undesirable

or unplanned), when any object or energy-interfering phenomenon scatters, or when standing wave reduces the transmission of light in an optical system; an attenuation will transmit something less than 100% of the light falling on its surface (incident).

Attenuator: An optical element which transmits some percentage of a laser beam away from the optical axis or point of incidence.

Back-Illuminated CCD: A CCD, also called "backside-illuminated CCD" and "back-thinned CCD", that is uniformly reduced to a thickness of approximately 10 μm so that an image can be focused on the back of the parallel register (where there is no gate structure); thinned CCDs exhibit a high sensitivity to photons, ranging from the soft X-ray to the near-infrared regions of the spectrum; since light is hitting the silicon directly, instead of passing through the gate structure, this type of CCD has excellent responsivity to blue light; many back-illuminated CCDs also have UV coatings that "down convert" UV light into the visible portion of the spectrum, further increasing the QE.

Backside Accumulation: A surface passivation technique required by backside-illuminated CCDs to achieve a high and stable QE.

Backside Charging: A QE accumulation technique to induce negative charge on the surface of a backside-illuminated CCD.

Backside Illumination: A CCD technology where incident photons enter the back of a sensor to achieve the highest possible QE.

Backside Well: A small potential well that develops at the surface of a backside-illuminated CCD after thinning.

Band-Gap Energy (E_G): The minimum energy that a valence electron must acquire to jump into the conduction band; a characteristic of the semiconductor or insulator material. Band-gap is described as a region between the valence band and the conduction band, devoid of allowed energy states.

Barrier Phase: A phase or region in a CCD that exhibits the lowest potential to confine the charge to a collecting phase or region.

Beam Splitter: An on-axis optical device to split, in varying percentages, a single beam into two beams; commonly, one beam is reflected from the beam splitter or one is transmitted through the beam splitter.

Bird's Beak: The interface between the gate oxide over the signal channel and the thicker field oxide over the channel stop.

Birefringent: A material whose refractive index changes according to the changing polarization states of incident light.

Blooming: A full-well condition of a CCD sensor where the charge escapes a collecting region through a barrier region.

Bond Strain Gradient (BSG) Model: The strained region in the SiO_2 layer near the Si-SiO_2 interface is due to a decrease in the Si-O-Si bond angle as the flexible SiO_2 network is forced to match with the silicon lattice.

Bulk (or Displacement) Damage: The damage associated with the displacement of silicon atoms from the lattice structure.

Bulk Dark Current: Dark current that is thermally generated in the bulk silicon.
Bulk Trap: A defect or impurity in the silicon wafer which traps signal electrons.
Buried-Channel CCD: A CCD technology where signal carriers are collected and transferred in a channel, located below and away from the Si-SiO$_2$ interface.
CCD (Charged-Coupled Device): A charge-coupled device is a light-sensitive integrated circuit that stores and displays the data for an image in such a way that each pixel (picture element) in the image is converted into an electrical charge, the intensity of which is related to a color in the color spectrum.
CCD Gates: Conductive electrodes that define the pixel boundaries and are blocked to collect and transfer the signal charge.
CCD Read-Out: CCDs are analog devices; to obtain a digital signal that is appropriate for doing a quantitative analysis, it is necessary to convert the analog signal to a digital format; when light is gathered on a CCD and is ready to be read out, a series of serial shifts and parallel shifts occurs. First, the rows are shifted in the serial direction towards the serial register. Once in the serial register, the data is shifted in the parallel direction out of the serial register, into the output node, and then into the analog-to-digital (A/D) converter where the analog data is converted into a digital signal.
Charge Capacity: The amount of charge that can be held by a pixel before blooming or surface interaction occurs
Charge Collection Efficiency (CCE): The efficiency of maintaining signal charge in the target pixel after a charge is generated.
Charge Generation Efficiency (CGE): The efficiency of the CCD to intercept incoming photons and generate electron-hole pairs; this efficiency is quantified by the quantum efficiency (QE).
Charge Transfer Efficiency (CTE): The fraction of charge successfully transferred per pixel transfer.
CMOS (Complementary Metal Oxide Semiconductors): Combination of both p-type channel and n-type channel MOS transistors on the same circuit design; CMOS circuits consume relatively low power.
Collection Phase: A high-potential phase that collects the signal electrons.
Color Centers: Absorbing sites in the lens element of a laser optical system, induced by repeated exposure to high intensity radiation.
Compton Effect: The process of photon scattering off loosely bound electrons in which the photon imparts a portion of its energy to the electron; the magnitude of the energy transfer depends on the scattering angle.
Conduction Band: An energy band in a solid in which electrons are freely mobile and can produce a net electric current.
Conduction Band Electron: An electron that is free to diffuse within the silicon lattice.

Current Density: The current per unit area.

Damage Threshold: A specific input energy level, usually radiant laser energy, which produces sufficient absorption in a lens or fiber optical material to cause solarization, color centers, or preablation sites; all of which reduce the transmission of the optical material and are considered as damage.

Dangling Bond: An unsatisfied bonding site at the Si-SiO$_2$ interface which creates traps.

Dark Signal Non-Uniformity (DSNU): The spatial variation of the dark signal within an image sensor, commonly expressed in terms of the differences in dark current from pixel to pixel, and specified as a percentage of the average signal.

Dark Current: Carriers that are thermally generated under completely dark conditions.

Dark Spikes: Isolated pixels that thermally generate dark current at a greater rate than the average dark current floor.

Deep Trap: A type of trap whose emission time constant is longer than the shortest clock overlap time period.

Deep-Depletion CCD: A custom CCD sensor, fabricated on high-resistivity bulk silicon, to extend the responsivity to the spectral range in the near-IR and hard X-ray regimes.

Deep-UV (DUV): The portion of the ultraviolet (UV) spectrum from approximately 180 nm to 280 nm, so-called because it is the deepest area of the UV where practical UV imaging is routinely done. It is also the deepest part of the UV spectrum, where irradiation work can be done in atmosphere conditions without side efforts. In this book, 157 nm is also considered as DUV.

Defect Cluster: A region where several radiation damage defects are closely grouped.

Delta Doping: An accumulation technique that grows a very thin highly doped epitaxial layer on the back surface of a CCD to obtain the QE-pinned condition.

Depletion Dark Current: The dark current generated thermally in the depletion region.

Depletion Region: The region in the CCD where dopant atoms are ionized by applied gate and channel voltages that generate a potential well.

Dielectric: A material having a relatively low electrical conductivity, such as an insulator; a substance that contains few or no free electrons (e.g., silicon dioxide (SiO$_2$) and silicon nitride).

Diffusion Dark Current: Dark current carriers that diffuse from regions outside the depletion region.

Displacement Damage: See "Bulk Damage".

Divacancy: A trap defect caused by two adjacent vacancies in the silicon lattice.

Dose: See "Radiation Absorbed Dose".

Dynamic Range: The ratio of the CCD signal at full-well condition to the read noise.

E' Center: A relaxed oxygen vacancy in SiO_2, represented as $\equiv Si^{\bullet}$; the E' center is an oxygen-related point defect and constitutes a trivalent silicon atom that has an unpaired electron in a dangling orbital and is back-bonded to three oxygen atoms.

Elastic Interaction: An elastic scattering interaction of a high-energy ion and the coulomb potential, presented by the target ion that can result in displacement damage.

Electromagnetic (EM) Spectrum: The entire range of radiation extending in frequency from approximately 10^{23} Hz to 0 Hz or, in corresponding wavelengths, from 10^{-13} cm to infinity and including, in order of decreasing frequency, cosmic-ray photons, gamma rays, X-rays, ultraviolet radiation, visible light, infrared radiation, microwaves, and radio waves.

Electron: The signal carrier generated, collected, transferred, and measured in n-channel CCDs.

Electron-Hole (e-h) Pair: A carrier pair, produced when a photon or particle photoelectrically interacts with a silicon atom.

Emission Time Constant: The time period for a signal electron to escape from a trap thermally.

Epitaxial Layer: A high-quality layer of silicon, grown on a substrate where all the CCD functions take place.

Epitaxial Silicon: A type of silicon wafer, used in making high-performance CCDs.

Excimer: An acronym for "excited dimmer"; a molecule consisting of generally two atoms (e.g., krypton and fluorine) which are strongly bound in the excited or upper energy level state, and dissociate in the ground state; an excited complex of two molecules.

Excimer Lasers: Chemical lasers that are capable of generating very short wavelength UV radiation; e.g., KrF (248 nm), ArF (193 nm), and F_2 (157 nm); commonly used as a source of UV radiation in very-high resolution photolithography.

Exposure Time: The length of time that a CCD is accumulating the charge.

Extreme Ultraviolet (EUV): The spectral range that covers wavelengths of 10 nm to 100 nm.

Field-Assisted Emission: A process by which electrons are emitted from a trap and accelerated by an electric field.

Field Oxide: A relatively thick oxide (typically 100 nm to 500 nm) formed to passivate and protect semiconductor surface outside of active device area, and to provide lateral isolation between adjacent device structures. The field oxide layer is common in MOS structures, but does not participate in device operation.

Fixed Charge: The type of charge in a Si-SiO_2 structure, located in the oxide layer in the immediate vicinity of Si surface; the fixed charge does not move and does not exchange charge with Si, but has an electrical influence on the characteristics of a Si-SiO$_2$ structure.

Fixed Pattern Noise (FPN): An image noise that results from sensitivity differences between the pixels; also called "pixel nonuniformity".

Flatband Voltage (V_{FB}): A voltage, in MOS devices, at which there is no electrical charge in the semiconductor and, therefore, no voltage drop across it; in the energy band diagram, the energy bands of the semiconductor are horizontal (flat).

Fluence: The total concentration of the irradiated particles that impinge on a device; given in particles/cm^2.

Fractional Yield: The fraction of holes that remain in the gate dielectric after initial recombination and electron migration, after being generated by ionizing radiation.

Frenkel Pair: The atom-vacancy pair formed when an atom is dislocated from its position in a lattice by high-energy particles.

Frontside Illumination: A CCD technology where incident photons enter on the gate side of the sensor.

Full Well: The maximum charge level that a pixel can hold and transfer; also referred to as the "well capacity".

Fused Silica: Silicon dioxide (SiO$_2$) that is highly purified; also known as "vitreous silica".

Gate Structure: The polysilicon structure that is located on the parallel register in a traditional CCD; polysilicon gates are transparent at long wavelengths, but become opaque at wavelengths shorter than 400 nm.

Glass: A supercooled liquid composed of silica (SiO$_2$) and impurities; highly purified silica glass will transmit UV.

Hard CCD: A CCD in which custom fabrication process steps are used to reduce the effects of radiation damage.

Hole: The signal carriers generated, collected, and measured in p-channel CCDs.

Hopping Conduction: A dark current generation process where valence electrons transit through interface and bulk states into the conduction band.

Horizontal Shift Register: The register responsible in shifting the signal charge horizontally to the output amplifier; also referred to as the "serial register".

Hydrogen Passivation: The process by which hydrogen atoms are introduced to a CCD during processing, to passivate dangling bonds at the Si-SiO$_2$ interface.

Impact Ionization: A process in which a high-energy electron interacts with the silicon lattice, breaking the Si-Si covalent bonds, and generating electron-hole pairs.

Indium Tin Oxide (ITO): A material used in some CCD gates to provide a higher QE, particularly in the blue-green region of the spectrum.

Inelastic Interaction: An interaction typically involving a high-energy particle, and a silicon nucleus that results in displacement damage.
Infrared (IR): Invisible radiation with wavelengths from $0.7\,\mu m$ to $15\,\mu m$.
Integrated Circuit (IC): A chip etched or imprinted with a network of electronic components such as transistors, diodes and resistors along with their interconnections.
Integration: The act of accumulating signals or charges on a CCD.
Interface State: Mid-band energy states found at the Si-SiO$_2$ interface that are responsible for dark current generation and charge trapping; also referred to as "interface traps".
Interstitial: The atom in an atom-vacancy pair, produced when an atom is dislocated from its position in a lattice; see also "Frenkel Pair".
Ionization Damage: The damage caused by the generation of electron-hole pairs within the gate dielectric.
Ionization: The process by which neutral atoms become electrically charged, either positively or negatively, by the loss or gain of electrons.
Ionizing Radiation: Radiation sources that produce ionizing particles that damage the gate dielectric of the CCD.
Laser: An acronym for "Light Amplification by the Stimulated Emission of Radiation"; a stimulated emission device that produces intense, highly coherent, monochromatic optical radiation.
Light Shield: A metal layer deposited over regions of the CCD to shield them from incoming light.
Lithography: The transfer of a pattern from one medium to another, for example, transferring a pattern from a mask or reticle to a wafer.
Local Oxidation of Silicon (LOCOS): Oxidation of selected areas of a silicon wafer by masking off the oxidation reaction from other regions. A thin uniform SiO$_2$ layer is initially formed (known as pad oxide) and then a layer of silicon nitride, Si$_3$N$_4$, is deposited. The Si$_3$N$_4$ is photolithographically patterned and then a relatively thick SiO$_2$ layer is grown in the openings in the Si$_3$N$_4$. Oxidation is inhibited at regions with Si$_3$N$_4$. Following oxidation the Si$_3$N$_4$ layer is stripped off the wafer. The thin pad oxide layer is used to help relief stress from direct contact between Si and Si$_3$N$_4$. LOCOS is widely used to isolated MOSFETs.
Lumogen: A phosphor coating, applied to the CCD, to extend the QE response into the UV and EUV spectral regimes.
Microlithography: The science of imaging micron and submicron structures onto silicon wafers and other substrates by using a photomask with patterns of the image and a photoresist coating on the substrate onto which the mask pattern is formed.
Micron: Short for micrometer (10^{-6} m), a unit of measure that is 1 millionth of a meter, or 1 thousandth of a millimeter.
Multi-Pinned Phase (MPP): A multiphase CCD technology that suppresses surface dark current generation.

MOS (Metal-Oxide-Semiconductor): A three layer structure in which the concentration of charge carriers in the semiconductor's sub-surface region (or the current flowing in semiconductor in the direction parallel to its surface) is controlled by the potential applied to a metal contact, or in other words, by the field-effect; the core of the MOS field-effect transistors (MOSFET).

MOSFET (Metal-Oxide-Semiconductor Field-Effect Transistor): A FET with a MOS structure as a gate, where the channel is created by inverting the semiconductor surface underneath the gate.

Mobile Charge: Electrically charged species which can move in the MOS gate oxide under the influence of an electric field, and can cause instability in the MOS device's characteristics; Na^+ ions are the most common mobile charges in SiO_2.

Non-Bridging Oxygen (NBO): Non-bridging oxygen refers to an oxygen atom that fails to bridge one SiO_4 tetrahedral to another tetrahedral in the SiO_2 network; the NBO can be represented by $\equiv Si - O^\bullet$, and is a source of intrinsic point defects; when the NBO captures a hole, it is referred to as an oxygen-hole center (OHC) or a non-bridging oxygen hole center (NBOHC, $\equiv Si - O^\bullet$).

Non-Bridging Oxygen Hole Center (NBOHC): Represented as $\equiv Si - O^\bullet$; see also "Non-Bridging Oxygen".

Nonlinear Absorption: The process wherein the attenuation coefficient becomes a function of the light intensity.

Open-Pinned-Phase (OPP): A high-QE frontside-illuminated CCD technology that has a portion of the pixel open to incident photons.

Output Amplifier: A MOSFET amplifier that provides an output voltage for each pixel.

Output Node: The location on a CCD (often a single pixel, adjacent to the serial register) where the charge is collected as a discrete picture element for a CCD read-out; the data enters the output node from the serial register and exits through the analog-to-digital (A/D) converter.

Output Transfer Gate (OTG): The last gate of the horizontal register which is used for the charge injection and the clock isolation between the sense node and output summing well.

Overflow Drain: A drain on the opposite edge of the device that keeps the thermally generated charge from entering the device; the overflow drain is used for CCDs with only one horizontal register.

Oxygen-Deficient Center (ODC): An oxygen vacancy which is classified into two variants: ODC(I) and ODC(II); the structural origin of ODC(I) in SiO_2 is attributed to the $\equiv Si–Si\equiv$ homobond; ODC(II) is assigned to an unrelaxed neutral oxygen vacancy which is less stable than the $\equiv Si–Si\equiv$ bonds.

Pair Production: A process in which a high-energy photon collides with a target atom and creates an electron-positron pair.

Phase-Shift Mask: A chrome-on-quartz or surface-etched quartz with patterns to shift the phase of the light in selected areas so as to permit a better overall patterning fidelity and to compensate for the proximity effects and other nonlinear geometry-related patterning effects.
Phosphor: A chemical substance that fluoresces when excited by X-rays, an electron beam, or UV radiation; phosphors are composed of rare earth oxides or halides (e.g., gadolinium, lanthanum, and yttrium) and usually emit green light with decay times that range from hundreds of nanoseconds to a few milliseconds.
Photo-Response Non-Uniformity (PRNU): The spatial variation of the photo-induced signal generating process within an image sensor.
Photoelectric Effect: A process in which valence band electrons are injected into the conduction band by a photon interaction.
Photoelectron: An electron that has been ejected from its parent atom by the interaction between that atom and a photon.
Photoemission: The process by which interacting photons stimulate the emission of electrons from one material into another material (e.g., the photoemission of electrons in the silicon layer into the SiO_2 layer).
Photolithography: The process of defining the polysilicon and gate dielectric layers and implants for CCD and MOS integrated circuits; light is used to transfer a pattern or image from one medium to another such as from a mask to a wafer; microlithography refers to the process that is applied to image with features in the micrometer range.
Photon: An elementary particle, or quantum, of radiant energy that stimulates the CCD and generates electron-hole pairs provided that the photon has sufficient energy.
Peroxy Linkage (POL): Represented as $\equiv Si - O - O - Si \equiv$, a defect in the SiO_2.
Peroxy Radical (PR): Represented as $\equiv Si - O - O^\bullet$, an excess-type defect in the SiO_2.
Pinning Implant: A highly concentrated, shallow boron p-type doped layer that pins the surface potential to the substrate potential.
Pinning: A bias condition that occurs when the signal channel is driven into inversion and pins the Si-SiO_2 surface potential to the substrate potential.
Pixel Nonuniformity: Variations in the pixel sensitivity to incident photons.
Pixel: A picture element; the smallest resolved area or object in a given image.
Point Defects: Highly localized imperfections of a crystalline structure, including vacancies, interstitials, and substitutional defects; point defects affect the periodicity of the crystal mostly in, or around, one unit cell.
Polysilicon: Highly doped, semitransparent, semiconductive, noncrystalline silicon used to form the gates of a CCD.
Potential Maximum: The highest potential within the potential well.

Potential Well: The potential distribution in the signal channel, which is responsible for the charge collection and the charge transfer.

Process Traps: Charge transfer efficiency (CTE) traps that result as the CCD is processed.

P-V Center: A phosphorus-vacancy center that traps a single electron induced when the CCD is exposed to high-energy particles.

Quantum Efficiency (QE): The number of electrons generated per incident photon.

Quantum Yield: The number of electrons generated per interacting photon.

Radiation Absorbed Dose (RAD): The standard unit of radiation dose, equivalent to the deposition of 100 ergs of energy per gram of material in the form of electron-hole pairs.

Radiation Events: Charge generated by energetic ions interacting with the CCD.

Radiation Hard: Describes devices that are able to withstand a higher-than-normal dose of ionizing radiation within the specified limits of performance degradation.

Radiation Shield: A shield, usually made of aluminum or tantalum, placed near a CCD to reduce the dose received by the CCD.

Radiation Traps: Charge transfer efficiency (CTE) traps induced when the CCD is exposed to high-energy particles.

Resolution: The smallest image that can be clearly discerned with the instrument and technique used, in terms of either space (spatial resolution), time (temporal resolution), or intensity.

Responsivity: The absolute QE given in units of amps per watt.

Reverse Annealing: The process by which an irradiated CCD's dark current and flat-band shift continue to increase long after the initial exposure.

Saturation: The absolute maximum signal level possible in the device; it is usually determined by factors such as the onset of clipping, (e.g., with anti-blooming), charge spreading (e.g., in non-antibloomed devices), or gross non-uniformity.

Scientific-Grade CCD: A high-performance CCD that offers fewer defects than commercial-grade CCDs. Scientific-grade CCDs produce better spatial resolution, have lower noise, and enable the user to accurately measure the intensity differences between the objects.

Sense Node: The region where the signal charge is dumped from the horizontal register, allowing the measurement of the charge packet size as a voltage; it is also called the "floating diffusion" or "output diode".

Shallow Trap: A type of trap whose emission time constant is shorter than the shortest clock overlap time period.

Silicon Dioxide (SiO_2) or Silica: Exists in crystalline or amorphous form, and occurs naturally in impure forms such as quartz and sand; in semiconductor technology, SiO_2 is used in the form of amorphous thin films, and is the most common insulator in MOS devices; very high quality films are obtained by the thermal oxidation of silicon which forms an excellent interface with the silicon; single crystal SiO_2 is known as quartz, and amorphous SiO_2 is sometimes referred to as glass.

Soft CCD: A CCD in which no special effort has been made in the design and the fabrication to lessen the effects of radiation damage on the performance.

Solid-State: Refers to the electronic properties of crystalline material, as opposed to vacuum and gas-filled tubes that transmit electricity; compared with earlier vacuum-tube devices, solid-state components are smaller, less expensive, more reliable, use less power, and generate less heat.

Surface Dark Current: Thermally generated dark charge produced at the Si-SiO_2 interface.

Surface-Channel CCD: An early CCD technology where a charge is collected and transferred at the Si-SiO_2 interface.

Thermal Anneal: A process by which the radiation traps are thermally annealed.

Thin Gate: A frontside-illuminated CCD technology that uses an ultrathin (<400 Å) gate as one of the phases for a high QE response.

Total Dose: The ionizing radiation dose received by a device. See also "radiation absorbed dose".

Transparent Gate: A frontside CCD technology that can deliver high QE, where one of the polygate electrodes is replaced by an optically transparent conducting gate material.

Trap: An undesired region in the signal channel where electrons are deferred.

Ultraviolet (UV): Wavelengths in the electromagnetic spectrum that begins at the end of the "violet" portion of the visible spectrum (400 nm), and extends down to where the X-ray region begins (10 nm).

UV Flooding: A backside charging technique used to accumulate and negatively charge the surface of a backside-illuminated CCD.

UV Lamp: A lamp that emits a sizable quantity of UV radiation; examples are mercury arc and xenon or deuterum lamps which are enclosed by a quartz envelope; the lamps can contain "dupont" gases to increase the spectral output in the UV region.

UV Laser: A laser such as an excimer laser, quadrupled Nd:YAG laser, or He:Ne laser, that emits radiation in the UV region.

Vacancy: A defect that is produced when a silicon atom is displaced from its position in the lattice; a missing ion in a lattice point.

Vacuum Ultraviolet (VUV): The region of the ultraviolet spectrum between 200 nm and the shortest end of the UV spectrum, about 10 nm; VUV radiation generally requires the use of a vacuum environment to prevent the absorption of the VUV energy by gas (air) molecules.

Valence Band: The highest occupied energy level in a solid crystal semiconductor or insulator that is occupied by electrons at $T = 0$ K.

Vertical Shift Register: The register responsible for shifting the signal charge vertically to the horizontal shift register; also referred to as "parallel register".

Via: A small opening in an insulative layer which is filled with a metallic conductor to permit an ohmic contact with the underlying silicon semiconductor device; multiple layers of semiconductor devices are connected with via metallization.

Virtual Phase CCD: A single-phase frontside-illumination CCD technology that exhibits high UV sensitivity.

Visible Light: The region of the electromagnetic spectrum, with wavelength from 400 nm to 700 nm, that can be perceived by the eye or the human retina.

Wafer: A semiconductor substrate, sliced from a crystalline ingot of silicon or gallium arsenide or sapphire, and polished on one side to an optical finish; the wafers are then cleaned, patterned with a resist layer, etched, and doped. After several such operations and the final metallization steps, the wafers are cut into individual dies.

Wavelength: The physical distance covered by one cycle of a sinusoidal wave of electromagnetic radiation; or the distance between the phase maxima or wave peaks in a light beam.

References

1. Duncan Technologies Inc., 2000.
   Available: http://www.duncantech.com/Dig_Camera_OV.htm
2. W. Franks, *Inorganic Phosphor Coatings for Ultraviolet Responsive Image Detectors*, MASc Thesis, E&CE, University of Waterloo, ON, Canada, 2000.
3. A.K. Bates, M. Rothschild, T.M. Bloomstein, T.H. Fedynyshyn, R.R. Kunz, V. Liberman, and M. Switkes, "Review of technology for 157-nm lithography," *IBM Journal of Research and Development*, Vol. 45, No. 5, pp. 605-614, 2001.
4. D. Lammers, "Industry weighs shift to 157-nm lithography," *EE Times*, May 2000. Available:
   http://www.eetimes.com/article/showArticle.jhtml?articleId=18304023
5. D. Lammers, "Lithography is wet and wild," *EE Times*, August 2003. Available:
   http://www.eetimes.com/article/showArticle.jhtml?articleId=18309155
6. D. Lammers, "Lithography choice: wet or dry?" *EE Times*, July 2003. Available:
   http://www.eetimes.com/article/showArticle.jhtml?articleId=18308826
7. "38-nm resolution immersion lithography demonstrated," *OE Magazine*, February 2004. Available:
   http://oemagazine.com/newscast/2004/020404_newscast01.html
8. "Front-end executive outlook," *Semiconductor International*, June 2004.
   Available: http://www.reed-electronics.com/semiconductor/article/CA425763
9. R. Dawson, R.M. Dawson, R. Andreas, J.T. Andrews, M. Bhaskaran, R. Farkas, D. Furst, S. Gershstein, M.S. Grygon, P.A. Levine, G.M. Meray, M. O'Neal, S.N. Perna, D. Proefrock, M. Reale, R. Soydan, T.M. Sudol, P.K. Swain, J.R. Tower, and P. Zanzucchi, "Deep UV sensitive high-frame-rate backside-illuminated CCD camera developments," *Proc. SPIE*, Vol. 4669, pp. 184-192, 2002.
10. R.A. Stern, R.C. Catura, R. Kimble, A.F. Davidsen, M. Winzenread, M.M. Blouke, R. Hayes, D.M. Walton, and J.L. Culhane, "Ultraviolet and extreme ultraviolet response of charge-coupled-device detectors," *Optical Engineering*, Vol. 26, No. 9, pp. 875-883, 1987.
11. "CCD Technology Primer," DALSA Corporation, Waterloo, ON, Canada, 1998.
12. A. Theuwissen, *Solid-State Imaging with Charge-Coupled Devices*, Kluwer Academic Publishers, Boston, 1995.
13. *Kodak CCD Primer*, Eastman Kodak Company – Microelectronics Technology Division, Rochester, New York, 1999.
14. J.R. Janesick, *Scientific Charge-Coupled Devices*, SPIE Press, Washington, 2001.

15. J. Buey, "First results on the Corot CCD test bench," France, 2003. Available: http://pccorot15.obspm.fr/COROT-CAL/Colloques/mons1.pdf
16. R. Widenhorn, M.M. Blouke, A. Weber, A. Rest, and E. Bodegom, "Temperature dependence of dark current in a CCD," *Proc. SPIE*, Vol. 4669, pp. 193-201, 2002.
17. A. Streitwieser, C. Heathcock, and E. Kosower, *Introduction to Inorganic Chemistry*, 4th Edition, Macmillan Publishing Company, USA, 1992.
18. D.J. Elliot, *Ultraviolet Laser Technology and Application*, Academic Press Inc., USA, 1995.
19. W.W. Duley, *UV Lasers: effects and applications in materials science*, Cambridge University Press, Cambridge, 1996.
20. M. Rothschild, T.M. Bloomstein, J.E. Curtin, D.K. Downs, T.H. Fedynyshyn, D.E. Hardy, R.R. Kunz, V. Liberman, J.H.C. Sedlacek, R. S. Uttaro, A. K. Bates, and C. Van Peski, "157 nm: Deepest deep-ultraviolet yet," *Journal of Vacuum Science & Technology B: Microelectronics and Nanometer Structures*, Vol. 17, No. 6, pp. 3262-3266, 1999.
21. K. Lewotsky, "Going the distance," *SPIE's oemagazine*, pp. 18-19, March 2002. Available: http://oemagazine.com/fromTheMagazine/mar02/specialfocus.html
22. M. LaPedus, "Intel revises litho roadmap amid 157-nm, EUV delays," *Silicon Strategies*, Feb. 2003. Available: http://www.siliconstrategies.com/
23. I. Lalovic, "DUV and VUV lithography application of excimer laser systems," *2001 Digest of the LEOS Summer Topical Meetings, IEEE*, pp. 15-16, 2001.
24. L. Peters, "Panel Agrees Immersion Lithography Will Happen, But When?" *Semiconductor International*, April 2004. Available: http://www.reed-electronics.com/semiconductor/article/CA405652
25. M.D. Whitfield, S.P. Lansley, O. Gaudin, R.D. McKeag, N. Rizvi, and R.B. Jackman, "Diamond photodetectors for next generation 157-nm deep-UV photolithography tools," *Diamond and Related Materials*, Vol. 10, pp. 693-697, 2001.
26. S. Stokowski and M. Vaez-Irvani, "Wafer Inspection Technology Challenges for ULSI Manufacturing," KLA-Tencor Corporation, California, 1999.
27. "Sarnoff Announces Deep Sub-Micron UV Light Inspection Camera Technology," *Sarnoff News and Press Releases*, Sarnoff Corporation, May 6, 2002. Available: http://www.sarnoff.com/news/index.asp?releaseID=83
28. Lambda Physik GmbH, Germany, 2003.
Available: http://www.lambdaphysik.com/
29. Roper Scientific, Inc., USA, 2000-2003.
Available: http://www.roperscientific.com/
30. S. Kubota, N. Eguchi, and H. Masuda, "DUV lasers applied to semiconductor inspection and optical disk mastering," *IEEE 2001 Digest of the LEOS Summer Topical Meetings – Advanced Semiconductor Lasers and Applications*, pp. 23-24, July 2001.
31. "Introduction to Multichannel Spectroscopy," Oriel Instruments, USA, 2001. Available: http://www.oriel.com/
32. M.A. Browne, "UV spectroscopic microscopy," *IEE Colloquium on New Microscopies in Medicine and Biology*, 1994.
33. "Ultraviolet Resonance Raman Spectroscopy," LOT-Oriel, Italy, 2003. Available: www.lot-oriel.com/pdf_it/all/andorapp_uv_raman_it.pdf
34. J. Kastner, "ACIS Back-Illuminated CCDs," December 1997.
Available: http://asc.harvard.edu/udocs/news_05/node11.html.

35. T.J. Jones, P.W. Deelman, S. T. Elliott, P.J. Grunthaner, R. Wilson, and S. Nikzad, "Thinned charge-coupled devices with flat focal planes for UV imaging," *Proc. SPIE*, Vol. 3965, pp. 148-156, 2000.
36. S. Nikzad, M.E. Hoenk, P.J. Grunthaner, R.W. Terhune, R.J. Grunthaner, R. Winzenread, M.M. Fattahi, and H.F. Tseng, "Delta-doped CCDs for enhanced UV performance," *Proc. SPIE*, Vol. 2278, pp. 138-146, 1994.
37. K. Vogler, I. Klaft, T. Schroeder, U. Stamm, K.R. Mann, O. Apel, C. Goerling, and U. Leinhos, "Long term test and characterization of optical components at 193 nm and 157 nm," *Proc. SPIE*, Vol. 4102, pp. 255-260, 2000.
38. B.N. Mukashev, "Ion-Radiation Induced Modification of Silicon and Development New Technology for Its Production," Academy of Sciences of the Republic of Kazakstan, May 1999. Available: http://www.sci.kz/~mukashev/summary.html
39. E.D. Palik (ed.), *Handbooks of Optical Constants of Solids I*, Academic Press, USA, 1998.
40. H. Föll, *Semiconductor*, University of Kiel, October 2003. Available: http://www.tf.uni-kiel.de/matwis/amat/semi_en/index.html.
41. G. Barbottin and A. Vapaille (editors), *Instabilities in Silicon Devices: Silicon Passivation and Related Instabilities*, Vol. 1 and 2, Elsevier Science Publishers, The Netherlands, 1986.
42. J.D. Plummer, M. Deal, and P. Griffin, *Silicon VLSI Technology: Fundamentals, Practice and Modeling*, Prentice Hall, USA, 2000.
43. H. Nishikawa, "Structures and Properties of Amorphous Silicon – Issues on the Reliability and Novel Applications," from *Silicon-Based Materials and Devices: Properties and Devices*, Vol. 2, edited by H.S. Nalwa, Academic Press, USA, 2001.
44. D.L. Griscom, "Self-trapped holes in amorphous silicon dioxide," *Physical Review B*, Vol. 40, No. 6, pp. 4224-4227, 1989.
45. D.L. Pulfrey and N.G. Tarr, *Introduction to Microelectronic Devices*, Prentice Hall, USA, 1989.
46. C.F. Cerofolin and L. Meda, *Physical Chemistry of, in and on Silicon*, Springer-Verlag, Berlin, Heidelberg, 1989.
47. F.J. Grunthaner, P.J. Grunthaner, and J. Maserjian, "Radiation-induced defects in SiO_2 as determined with XPS", *IEEE Transactions on Nuclear Science*, Vol. NS-29, No. 6, pp. 1462-1666, 1982.
48. H. Kobayashi, A. Asano, J. Ivanco, M. Takahashi, and Y. Nishioka, "New spectroscopic method for the observation of semiconductor interface states and its application to MOS structure," *Acta Physica Slovaca*, Vol. 50, No. 4, pp. 461-475, 2000.
49. T.P. Ma and P.V. Dressendorfer (editors), *Ionizing Radiation Effects in MOS Devices and Circuits*, John Wiley & Sons, Inc., USA, 1989.
50. E.M. Young and W.A. Tiller, "Ultraviolet light stimulated thermal oxidation of silicon," *Applied Physics Letters*, Vol. 50, No. 2, pp. 80-82, 1987.
51. R. Williams, "Photoemission of electrons from silicon into silicon dioxide," *Physical Review*, Vol. 140, No. 2A, pp. 569-575, 1965.
52. A. Goodman, "Photoemission of holes from silicon into silicon dioxide," *Physical Review*, Vol. 152, No. 2, pp. 780-784, 1966.
53. K. Kajihara, Y. Ikuta, M. Hirano, and H. Hosono, "Power dependence of defect formation in SiO_2 glass by F_2 laser irradiation," *Applied Physics Letters*, Vol. 81, No. 17, pp. 3164-3166, 2002.

54. N.F. Borrelli, C. Smith, D. Allan, and T.P. Seward III, "Densification of fused silica under 193-nm excitation," *Journal of Optical Society of America B*, Vol. 14, No. 7, pp. 1606-1615, 1997.
55. J. Moll, "Assessing damage for UV-laser-resistant fused silica," *Photonics Spectra*, April 2002, pp. 78-81.
56. R.J. Araujo, N.F. Borrelli, and C. Smith, "Induced absorption in silica (a preliminary model)," *Proc. SPIE*, Vol. 3424, pp. 2-9, 1998.
57. Y. Ikuta, S. Kikugawa, M. Hirano, and H. Hosono, "Damage behavior of SiO_2 glass induced by 193 nm radiation under simulated operating mode of lithography laser," *Proc. SPIE*, Vol. 4347, pp. 187-194, 2001.
58. H. Hosono, Y. Ikuta, T. Kinoshita, K. Kajihara, and M. Hirano, "Physical disorder and optical properties in vacuum ultraviolet region of amorphous SiO_2," *Physical Review Letters*, Vol. 87, No. 17, pp. 175501/1-4, 2001.
59. H. Imai, K. Arai, H. Hosono, Y. Abe, T. Arai, and H. Imagawa, "Dependence of defects induced by excimer laser on intrinsic structural defects in synthetic silica glasses," *Physical Review B*, Vol. 44, No. 10, pp. 4812-4818, 1991.
60. K. Kajihara, L. Skuja, M. Hirano, and H. Hosono, "Formation and decay of nonbridging oxygen hole centers in SiO_2 glasses induced by F_2 laser irradiation: In situ observation using a pump and probe technique," *Applied Physics Letters*, Vol. 79, No. 12, pp. 1757-1759, 2001.
61. K. Arai, H. Imai, H. Hosono, Y. Abe, and H. Imagawa, "Two-photon processes in defect formation by excimer lasers in synthetic silica glass," *Applied Physics Letters*, Vol.53, No. 20, pp. 1891-1893, 1988.
62. M. Mizuguchi, L. Skuja, H. Hosono, and T. Ogawa, "Photochemical processes induced by 157-nm light in H_2-impregnated glassy SiO_2:OH," *Optics Letters*, Vol. 24 No. 13, pp. 863-865, 1999.
63. C. Fiori, R. Devine, and P. Meilland, "Photoinduced fixed oxide charge modification in SiO_2 films," *Journal of Applied Physics*, Vol. 58, No. 2, pp. 1058-1060, 1985.
64. M.W. Lee and C.K. Song, "Oxygen Plasma Effects on Performance of Pentacene Thin Film Transistor," *Japanese Journal of Applied Physics*, Vol. 42, Part 1, No. 7A, pp. 4218-4221, 2003.
65. Y.F. Lu, W.K. Choi, Y. Aoyagi, A. Kinomura, and K. Fujii, "Controllable laser-induced periodic structures at silicon-dioxide/silicon interface by excimer laser irradiation," *Journal of Applied Physics*, Vol. 80, No. 12, pp. 7052-7056, 1996.
66. K. Kurosawa, Y. Takigawa, Y. Takigawa, W. Sasaki, M. Katto, and Y. Inoue, "Vacuum ultraviolet laser-induced surface alternation of SiO_2," *Japanese Journal of Applied Physics*, Vol. 30, No. 11B, pp. 3219-3222, 1991.
67. L.R. Canfield, J. Kerner, and R. Korde, "Stability and quantum efficiency performance of silicon photodiode detectors in the far ultraviolet," *Applied Optics*, Vol. 28, No. 18, pp. 3940-3943, 1989.
68. C. Fiori and R. Devine, "Photo-induced oxygen loss in thin SiO_2 films," *Physical Review Letters*, Vol. 52, No. 23, pp. 2081-2083, 1984.
69. D.E. Husk, C. Tarrio, E.L. Benitez, and S.E. Schnatterly, "Response of photodiodes in vacuum ultraviolet," *Journal of Applied Physics*, Vol. 70, No. 6, pp. 3338-3344, 1991.
70. R. Korde and L.R. Canfield, "Silicon photodiodes with stable, near-theoretical quantum efficiency in the soft x-ray region," *Proc. SPIE*, Vol. 1140, pp. 126-132, 1989.

71. R. Korde, J. Cable, and L.R. Canfield, "One gigarad passivating nitrided oxides for 100% internal quantum efficiency silicon photodiodes," *IEEE Transactions on Nuclear Science*, Vol. 40, No. 6, pp. 1655-1659, 1993.
72. R. Korde and J. Geist, "Quantum efficiency stability of silicon photodiodes," *Applied Optics*, Vol. 26, No. 24, pp. 5284-5290, 1987.
73. S. Yoon and M. White, "Study of thin gate oxides grown in an ultra-dry/clean triple-wall oxidation furnace system," *Journal of Electronic Materials*, Vol. 19, No. 5, pp. 487-493, 1990.
74. International Radiation Detectors (IRD) Inc., Torrance, CA, 2002. Available: http://www.ird-inc.com/

Index

Absorption
  coefficient 12
  depth 12
  photon 11
  silicon 13, 24, 45
  silicon dioxide 54
Annealing 107, 114, 148
Anti-reflection coating 10, 16, 36

CCD 7
  157 nm measurement 157
  area array sensor 8
  back-illuminated 10, 16, 35, 42, 112
  deep-depletion 39
  DUV applications 27
  DUV challenges 29, 40
  DUV degradation 41
  front-illuminated 10, 16
    frontside-thinned 159
  ionization damage 110, 111
  linear sensor 7, 159
  open-electrode 34
  operation 7
  photodiode 7
  photogate 8
  radiation effects 109
  structure 7
  UV phosphor 38, 112
  virtual-phase 33
CCE 20, 50, 186
Charge collection 13
Color center 76
  DUV laser 133
  UV-induced 127
Compton scattering 99
CTE 50

Dark current 18, 166, 178

dark signal non-uniformity (DSNU) 19
  origins 18
Displacement damage 96
DUV
  CCD degradation 41
  definition 25
  excimer lasers 25
  imaging and CCD 27
    challenges 40
    laser beam profiler 31
    lithography 27
    microscope 31
    spectroscopy 32
    wafer inspection 30

E' center 62, 65, 76, 77, 100, 127, 131, 133, 136
Electromagnetic (EM) spectrum 23
Excimer laser 25, 125, 157
  applications 26
  DUV wavelengths 25
  lithography 2, 28

Fluorine doping 140, 206
Frenkel defects 97

Ionization damage 96
  CCD 110, 111
    annealing 114
    fractional yield 113
    hole transport 113
    hole trapping 113
    interface charge 97
    interface state 114
    oxide charge 97
    oxide thickness 116
    UV flood 117

Index

Ionization energy 96

Lithography 2, 27
  157 nm 2, 28
  CCD 29

MOSFET 68
  defects 77
  radiation effects 103

NBOHC 61, 66, 76, 100, 127, 132, 134, 136
Nitridation 204

Output node 20
  floating diffusion 14, 20
  floating gate 21
Oxygen deficient center (ODC) 130, 137

Peroxy linkage (POL) 65, 130
Peroxy radical (PR) 65, 76, 100
Photoelectric effect 7, 45, 99
Pixel response non-uniformity (PRNU) 17, 160
Point defects
  silicon 48
  silicon dioxide 57

Quantum efficiency 15, 181
  157 nm 162
  definition 16
  dispersion 17
  DUV fluctuation 163, 170
  extrinsic 15
  intrinsic 15
  UV photodiode 200
  visible 16

Radiation effects 95
  CCD 109
  MOS 103
Radiation hardness 98, 107

Si-SiO$_2$ interface
  dark current 18
  defects 82
    electrically-active 90
    P_b center 82, 88, 90
    radiation-induced 102
  interface charge 90
  interface states 83, 86
  physical structure 81
  UV-induced effects 153
Silicon 45
  absorption 13, 45
  defects 48
    radiation-induced 100
  epitaxial 49
  ionization energy 96
  optical properties 45
  photoelectric effect 45
  polycrystalline 1, 10, 41
  UV-induced effects 121
  wafer 49
Silicon dioxide
  absorption 54
  amorphous 52, 53
  broken bond (dangling bond) 59
  chemical disorder 129
  crystalline 52
  defects 52, 54
    electrically-active 56, 67, 76
    formation reactions 65
    optically-active 56, 75
    radiation-induced 100
  ionization energy 96
  non-bridging oxygen 61
  optical properties 54
  oxide charges 67, 78
    radiation-induced 101
  physical disorder 129
  strained bond 58, 129, 131, 139
  structural properties 51
  trapping process 70
  traps 69, 77
  UV transmission 54
  UV-induced effects 125
SiO$_2$ or silica *see* silicon dioxide
Spectral response 14

Two-photon absorption process 126, 171

UV
  definitions 24
  illumination sources 25
  laser 25
  photoemission 112, 121

Si-SiO$_2$ interface   153
silicon   121
silicon dioxide   125
UV laser effects   126
  bleaching   145, 173
  color center formation   133
  defect formation   139
  densification   128
  induced absorption   127, 134, 172, 176, 177
  interface modification   153, 173, 175
  oxide charging   147, 170, 173, 175
  photorefractive effect   129
UV photodiode
  157 nm measurement   188
  challenges   197
  enhancement techniques   198

Wafer inspection   30

Printing: Krips bv, Meppel
Binding: Litges & Dopf, Heppenheim